THE ECONOMICS OF ASYMMETRIC

Also by Brian Hillier

Macroeconomics: Models, Debates and Developments
The Macroeconomic Debate: Models of the Closed and Open Economy

THE ECONOMICS OF ASYMMETRIC INFORMATION

Brian Hillier

palgrave
macmillan

© Brian Hillier 1997

All rights reserved. No reproduction, copy or transmission of this publication may be made without written permission.

No paragraph of this publication may be reproduced, copied or transmitted save with written permission or in accordance with the provisions of the Copyright, Designs and Patents Act 1988, or under the terms of any licence permitting limited copying issued by the Copyright Licensing Agency, 90 Tottenham Court Road, London W1P 9HE.

Any person who does any unauthorised act in relation to this publication may be liable to criminal prosecution and civil claims for damages.

The author has asserted his rights to be identified as the author of this work in accordance with the Copyright, Designs and Patents Act 1988.

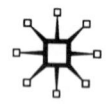

First published 1997 by
PALGRAVE MACMILLAN
Houndmills, Basingstoke, Hampshire RG21 6XS
and London
Companies and representatives
throughout the world

ISBN-13: 978-0-333-64749-3 hardcover
ISBN-10: 0-333-64749-1   hardcover
ISBN-13: 978-0-333-64750-9 paperback
ISBN-10: 0-333-64750-5   paperback

A catalogue record for this book is available from the British Library.

This book is printed on paper suitable for recycling and made from fully managed and sustained forest sources. Logging, pulping and manufacturing processes are expected to conform to the environmental regulations of the country of origin

10 9 8
10 09 08

Printed in China

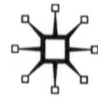

Published in the United States of America by
ST. MARTIN'S PRESS, INC.,
Scholarly and Reference Division
175 Fifth Avenue, New York, N.Y. 10010

ISBN 0-312-16396-7

To my children
Michael and Victoria (Vicky Sox)

Contents

List of Figures   x
Preface   xii

    Introduction   xiv

        The wisdom of Solomon   xiv
        The organisation of topics   xv

PART I   INVESTMENT FINANCE AND ASYMMETRIC INFORMATION

1   Asymmetric Information in the Market for Investment Finance   **3**
    1.1   Overview   3
    1.2   The selection problem   3
    1.3   The hidden action problem   4
    1.4   The costly state verification problem   4
    1.5   The agency problem   5

2   Investment Finance and the Selection Problem   **7**
    2.1   Overview   7
    2.2   The selection problem and the credit market   7
    2.3   The selection problem and equity finance   16
    2.4   Credit rationing   22
    2.5   Discussion   32
    2.6   Recommended reading   33
    2.7   Problems   34

3   Investment Finance and the Hidden Action Problem   **36**
    3.1   Overview   36
    3.2   Hidden action and the credit market   36
    3.3   The hidden action problem and equity finance   42
    3.4   Market collapse   44
    3.5   Credit rationing   50
    3.6   Problem   56

viii Contents

4	Investment Finance and the Costly State Verification Problem	57
	4.1 Overview	57
	4.2 Hidden information and the credit market	57
	4.3 The credit market and business cycles	67
	4.4 Recommended reading	74
	4.5 Problems	74

PART II ASYMMETRIC INFORMATION PROBLEMS IN THE INSURANCE MARKET

5	Insurance and Risk Aversion	77
	5.1 Overview	77
	5.2 Attitudes towards risk	77
	5.3 Risk aversion and insurance	80
	5.4 Recommended reading	88
	5.5 Problems	89

6	Insurance and the Hidden Action Problem	90
	6.1 Overview	90
	6.2 Insurance and the hidden action problem	90
	6.3 Recommended reading	96
	6.4 Problems	96

7	Insurance and the Selection Problem	98
	7.1 Overview	98
	7.2 Insurance and different risk categories under full information	99
	7.3 Pooling together different risk categories	100
	7.4 Separating contracts and equilibrium concepts	104
	7.5 Recommended reading	110
	7.6 Problem	111

PART III THE LABOUR MARKET: EDUCATION, SIGNALLING, SCREENING AND EFFICIENCY WAGES

8	The Selection Problem and Education	115
	8.1 Overview	115
	8.2 Education and screening	115
	8.3 Education and signalling	122
	8.4 Discussion	125
	8.5 Recommended reading	126
	8.6 Problems	127

9	The Hidden Action Problem and Efficiency Wages	128
	9.1 Overview	128
	9.2 Reasons for paying efficiency wages	129
	9.3 The hidden action problem and the shirking model	132

	9.4	Recommended reading	135
	9.5	Problems	135

PART IV REGULATION, PUBLIC PROCUREMENT AND AUCTIONS

10 Regulation and Procurement — **139**
    10.1 Overview — 139
    10.2 Regulation and hidden information — 139
    10.3 Procurement with hidden information and hidden action — 143
    10.4 Recommended reading — 151
    10.5 Problem — 151

11 Auctions — **153**
    11.1 Overview — 153
    11.2 Auctions and information problems — 154
    11.3 Private value auctions and the revenue equivalence theorem — 153
    11.4 Optimal auctions — 165
    11.5 Common value auctions and the winner's curse — 171
    11.6 Recommended reading — 172
    11.7 Problems — 173

Notes — 175
Bibliography — 182
Index — 184

List of Figures

2.1	The demand for loans	11
2.2	The relationship between ρ and r	13
2.3	The credit market and the selection problem	14
2.4	The equity market	21
2.5	The supply of deposits	23
2.6	The supply of loans	23
2.7	The credit market and the selection problem	25
2.8	The demand for shares	29
2.9	The equity market	29
2.10	The credit market with a backward-bending supply of loans curve	34
3.1	The demand for loans	38
3.2	Credit marker equilibrium	40
3.3	Equity markets under full information	46
3.4	Collapse of the share market	48
3.5	The supply of loans	50
3.6	The credit market and the hidden action problem	52
4.1	A uniform density function	58
4.2	Different share contracts	63
4.3	Hidden information and the credit market	66
4.4	Investment cost and expected excess returns under full information	69
4.5	The impact of costly state verification	70
4.6	Macroeconomic equilibrium	72
4.7	Persistence effects of shocks	73
5.1	Attitudes towards risk and the marginal utility of wealth	78
5.2	Risk aversion and insurance	81
5.3	The superiority of full insurance	84
5.4	The state–space representation	85
6.1	No full insurance at fair odds	91
6.2	Partial insurance at unfair odds	93
6.3	Hidden action and partial insurance	93
7.1	Insurance and different risk groups	99
7.2	The market average fair odds line	101
7.3	The pooling contract	102
7.4	Adverse selection	103

7.5	No Nash equilibrium under pooling	104
7.6	A pair of separating contracts	106
7.7	Separating contracts and Nash equilibrium	107
8.1	The full information case	116
8.2	Separating contracts and Nash equilibrium	119
8.3	Signalling and education	123
9.1	The no-shirking constraint	133
9.2	Equilibrium unemployment	134
10.1	Regulation under full information	140
10.2	Regulation under asymmetric information	141
10.3	Cost-plus contracts	145
11.1	Truth-telling and participation constraints	168
11.2	The probability constraints	169

Preface

Since the 1970s there have been rapid developments in the economics of asymmetric information. This sphere of economics deals with situations where agents on one side of the market know something that agents on the other side do not: for example, a seller of a second-hand car may have knowledge of its qualities unknown to a potential buyer. Such situations are very different from those dealt with in the more conventional analysis which assumes that buyers and sellers have the same information about goods being sold. On reflection, however, situations of asymmetric information seem to be widely prevalent in the real world, so that moving beyond the conventional analysis yields fascinating and handsome rewards.

This book makes the economics of asymmetric information accessible to students at an intermediate to advanced undergraduate level. It is also hoped that graduate students and others looking for a readable introduction to the area will find this text to be interesting. Furthermore, since the book is organised in terms of individual markets and topics such as auctions, it is hoped that it will prove useful as a reference or revision source for anyone interested in those specific areas. The terms 'he' and 'his' are used throughout the book only to avoid the clumsier alternatives, 'his/her' and 'she/he' and to render the arguments easier to follow. There is no intention to offend.

The level of mathematical sophistication required of the reader is kept to a minimum, and much use is made of verbal reasoning and diagrammatic analysis. The intention is to take the reader beyond the level of coverage possible in an intermediate microeconomics text, such as Varian (1992), which has an excellent chapter on information, without requiring a level of mathematical expertise beyond that of an intermediate-level student.

Much of the material in the book has been tried and tested for a number of years at the University of Liverpool, where Tim Worrall and I taught a third year microeconomics course and offered an option on Information Economics to Master-level students. In our experience, students take naturally to the strategic implications of behaviour under asymmetric information and the course was well received.

Our course was of the standard length at Liverpool; that is, about twenty lectures plus half a dozen tutorials. We did not cover all the material included in this book in the course: we omitted a couple of topics each year in order to fit the timetable. I recommend that anyone considering using the book as a text for a course, or part of a course, should cover the first four chapters to introduce the key issues of asymmetric information, cover Chapters 5–7 to introduce risk aversion (unless the students have already grasped these ideas from an earlier microeconomics course), and then pick and choose from the remaining chapters according to their own timetable and interests.

I thank Tim Worrall, now at the University of Keele, for his help and patience as we discussed how best to present certain topics in the classroom as well as in the book. I must also thank Murad Ibrahimo, whose Ph.D. work at the University of York, with which I was involved as supervisor before moving from York to Liverpool, first kindled my interest in the economics of asymmetric information. Many of the students on the course at Liverpool also deserve my thanks; in particular, I thank Martin Bayntun, Steve Brice, Jon Gershlick and Ben Sanderson. Last, but not least, I thank Jane Powell, the Commissioning Editor at Macmillan: I hope she has found a good solution to her selection problem and commissioned a good book.

BRIAN HILLIER

Introduction

The wisdom of Solomon

A good introduction to the idea of asymmetric information is provided by the well-known story of the wisdom of Solomon, as told in I Kings, 3. In this story, King Solomon, in a dream, has his wish for 'an understanding heart to judge thy people, that I may discern between good and bad' granted by God. Solomon's wisdom is illustrated by a story concerning two women who appear before him seeking judgement.

The women are described as two harlots, or prostitutes, who live together in a house. The women have a young baby with them and each claims to be its mother. Each woman claims that although the other woman also gave birth to a baby it died in the night, and that the other woman is the mother of the dead child and not of the living one. Solomon responds to the women's contradictory claims by instructing a servant to bring him a sword and to divide the child in half.

On hearing Solomon's command, the true mother of the child responds by saying that she is not the mother of the child and that it should be given to the other woman. The other woman says that the child should indeed be divided in two. Solomon is then able to tell who is the real mother and instructs that the child should not be killed but given to the true mother, who was prepared to give it away rather than see it die.

This story illustrates a number of points which recur throughout this book. First, there is a clear asymmetry of information; the women know whose baby the child is, but Solomon does not. Second, there is a conflict of objectives; Solomon would like to have the information that is available to the women in order to better achieve his goals. Third, the true mother would like to transmit the information but cannot easily do so because of the actions of the other woman, who also lays claim to the child. Finally, Solomon devises a contract to offer the women, which causes them to reveal their information to him.

We shall see in what follows that these points are mirrored in many market situations. Consider, for example, an insurance company offering accident insurance to car drivers, some of whom are naturally more cautious and less accident-prone than others. The insurance company is like Solomon because it cannot tell who is a safe driver and who is a risky one, but it would like to be able to do so in order to charge higher premiums to riskier drivers. The drivers are like the women, since both safe and risky drivers will claim to be safe to try to obtain cheaper insurance. Thus the insurance company, like Solomon, has to try to devise a contract to offer the drivers which will cause them to reveal themselves truthfully. Unlike Solomon, however, the insurance company is unlikely to be able to solve its problem perfectly; we shall see that the presence of the risky drivers prevents the safer ones from getting as good an insurance deal as they would otherwise be offered. An element in the story of Solomon which will not be so common in what follows is that he changes his mind and does not go ahead with cutting the child in two, although a similar point will recur in Chapter 10 below.[1]

The organisation of topics

Analysing situations of asymmetric information requires two significant departures from conventional analysis. One departure involves recognising and modelling the various types of asymmetry which appear in the literature, and the second involves seeing how the asymmetry affects the nature of the contract entered into by the participants in the market.

One obvious way to proceed is to introduce a type of asymmetry of information and examine how it operates in various markets and how it affects contracts in each market. The trouble with this approach, however, is that it is 'bitty'; it involves jumping from market to market within a chapter, while to get a picture of the impact of asymmetric information on a given market, for insurance, say, would require looking at several sections within a number of different chapters. We therefore follow an alternative approach of dealing with one market at a time, beginning with the market for investment finance. Our approach has the advantage that the reader may find an overview of how asymmetric information has an impact upon a particular market by looking at one chapter or a few consecutive chapters on that market, thus providing a useful introductory reference source for information problems in the markets covered in the book.

New terminology and techniques are introduced gradually as the analysis proceeds, and the reader new to this type of analysis is recommended to read the first seven chapters of the book consecutively. In particular, the ideas presented in Chapters 1–4 are drawn on throughout the remaining chapters and may be unfamiliar to readers. The last four chapters of the book use the ideas developed in the first seven chapters, and may be read in any order.

Most chapters end with some recommended reading and some problems. The recommended reading presents key articles that the reader might like to check to see how the ideas were developed in the original literature, or to delve deeper into any particular topic. The reader's appreciation of the ideas will be enhanced if he or she looks at some of the references given. The problems are designed to test the reader's understanding of the material covered and the reader will benefit from tackling them. Throughout the book I have tried not to avoid covering difficult ideas, and have tried to handle them in a simplified and accessible way. On the other hand, I have attempted to maintain the flow of the argument even if at times this has meant omitting a fairly interesting or relevant implication or extension. In some cases, the problems cover such points.

Part I
Investment Finance and Asymmetric Information

Part I

CHAPTER 1

Asymmetric Information in the Market for Investment Finance

1.1 Overview

Throughout this book we shall be examining three major categories of asymmetry of information and each of them is introduced in Part I. Here we imagine entrepreneurs wishing to raise funds in order to finance their investment projects. In the following chapters we discuss each of the categories of asymmetric information that might arise in this market. Although our treatment deals with each category separately, it is possible for them to occur simultaneously in the real world.

By the end of Part I of the book the reader should be aware of the categories of asymmetric information, how markets may respond to them, and how they can create problems not present in the usual symmetric information model. The analysis of much of the rest of the book will be based on some of the lessons and techniques introduced here. Throughout Part I of the book we avoid the complications of risk aversion by assuming that both entrepreneurs and suppliers of funds are risk neutral. The topic of risk aversion is introduced in Part II, which deals with the market for insurance.

Before going into a detailed examination of each of the categories of asymmetric information in the market for investment funds, it will be worthwhile to outline each of them briefly below. The final section of this chapter introduces some terminology.

1.2 The selection problem

We shall call the first type of asymmetric information problem to be introduced the *selection problem*. In this case we can imagine each entrepreneur to be endowed with one project in which he can invest. For simplicity, assume it to be common knowledge that all projects cost the same amount, K, and that entrepreneurs have no funds of their own. Hence, each entrepreneur needs to raise, by borrowing or by some other means, the entire amount K before he can fund his project.

Consider the return to any entrepreneur's project to be a random variable, R, drawn from some probability distribution. If the probability distributions from which project returns are drawn are different for different entrepreneurs, and if each entrepreneur alone knows the distribution from which the return will be drawn, the suppliers of funds may be said to face a selection problem. The problem for the suppliers of funds is that they may prefer to provide funds to entrepreneurs with some types of projects rather than others, but do not have the information to know which entrepreneurs have which projects.

This problem is sometimes said to be a problem of *ex ante* asymmetric information, since the asymmetry exists before the parties involved enter into an agreement with one another. Clearly, similar problems can arise in other markets; for instance, consider the example introduced earlier, where an insurance company wishes to insure careful drivers rather than careless ones, but is unable to tell one type of driver from another.

1.3 The hidden action problem

We shall call the second type of information problem to be introduced the *hidden action problem*. In this case, we can imagine each entrepreneur being able to choose from several different investment projects, the return for each being a random variable drawn from a different probability distribution. The suppliers of funds may be said to face a hidden action problem if they are unable to observe in which project the entrepreneur invests funds made available, and yet would prefer to fund some types of project rather than others. The problem is very similar to the selection problem, but here the suppliers of funds cannot observe the investment choice made by entrepreneurs after the funds have been made available to them, rather than facing the selection problem of being unable to distinguish between different types of entrepreneur.

This problem is sometimes said to be a problem of *ex post* asymmetric information, since the asymmetry occurs after the parties involved enter into an agreement with one another. Similar problems can arise in other markets; for example, a firm hiring a worker may not be able to observe the level of effort expended by the worker when at work.

1.4 The costly state verification problem

We shall call the third type of information problem to be introduced the *costly state verification problem*. In this case we can imagine each entrepreneur to be endowed with one project. The return to each project is a random variable and all entrepreneurs who invest receive an independent return drawn from the same probability distribution. In this case we can also assume that the suppliers of funds know the probability distribution from which project returns will be drawn and can observe the act of investment by the entrepreneur, thus the suppliers of funds face no problem of selection or hidden action.

In the two previous cases we have assumed implicitly that once a project yields a return, that return is observed by both the entrepreneur and the suppliers of funds, but now we assume, instead, that the entrepreneur alone freely observes the return yielded by his project. An entrepreneur may have an incentive to lie about his return to the suppliers of funds (for example, he may claim that he is unable to pay them anything since the project return was zero, even though this may be a lie). Since it is usual to assume that the suppliers of funds can observe the project yield by incurring a cost (the cost of sending in auditors to the entrepreneur's firm, say), this problem is known as the costly state verification problem. The jargon arises because in statistical terms the project yield, or outcome yielded by the drawing from the probability distribution, is known as the *state of the world*.[1]

Like the hidden action problem, this problem is sometimes said to be a problem of *ex post* asymmetric information, since the asymmetry occurs *after* the parties involved enter into an agreement with one another. This problem can also arise in other markets; for example, an employer may ask a worker to carry out a task but the level of difficulty of that task may not be apparent until the worker has begun it, and even then the difficulty may be apparent only to the worker and not to the employer. Another example is self-assessment tax regimes, where taxpayers may have an incentive to lie about their tax liability.

1.5 The agency problem

The types of problem we have introduced above are often called *agency* or *principal–agent problems*. In these problems the *principal* is the *uninformed* party and the *agent* is the *informed* party.

We assume throughout that agents are purely self-interested and are willing to lie to the principal about the information they have but which the principal does not have, whether they have this information prior to signing a contract with the principal or acquire it after signing the contract. Similarly, we assume that the agents are willing to deceive the principal about their actions if they perceive it to be to their advantage to do so. Knowing that agents behave in this way, the principal takes into account the behaviour of the agents when deciding which contract to offer them. Thus we model the interactions between the principal and agents as the interactions between players in a game and will use some of the techniques and terminology of *game theory*. Our emphasis will, however, be on the economic situation facing the players in the game and not on the mathematical technicalities of game theory, which we keep to a strict minimum.[2]

The emphasis on self-interest and the willingness to cheat is not meant to deny that, at times, individuals in the real world take altruistic actions, or to imply that people are all liars and cheats. Nevertheless, I would argue that the approach adopted here offers a reasonably accurate description of many real-world scenarios.

CHAPTER 2

Investment Finance and the Selection Problem

2.1 Overview

In this chapter we examine the implications of the selection problem in the market for investment finance. We begin in section 2.2 by assuming that the only way that entrepreneurs can raise finance for their projects is by borrowing from banks in a competitive credit market. This assumption allows us to ignore an important element in the analysis of an asymmetric information problem; that is, the form of contract offered by the principal to the agent. Section 2.2 shows that the selection problem in the credit market can lead to problems, including a socially inefficient level of investment. Section 2.3 considers whether it is possible to find a better form of contract than credit to provide funds for investment, and shows that the selection problem present in section 2.2 can be solved by the use of equity finance. Section 2.4 shows how the selection problem of the type presented in section 2.2 can lead to rationing in the credit market, which is an interesting result, since rationing is difficult to derive using conventional analysis. Finally, section 2.5 discusses a variety of the issues raised and possible responses to them. The chapter closes with recommended reading and some problems.

2.2 The selection problem and the credit market

Project returns

In this section we illustrate the selection problem by examining a specific example in which each entrepreneur has one project in which he can invest. All projects cost K and each entrepreneur needs to raise the entire amount K if he is to fund his project. This latter assumption rules out any complications provided by the possibility of the entrepreneur investing some of his own wealth in the project, although we shall consider relaxing this assumption in section 2.5 below.

8   Investment Finance

The return to a project is the amount R_i, which is a random variable, where the subscript i represents the ith project. Assume that any project either succeeds in the period after it has been set up–the ith project yielding the project specific return R_i^s with probability p_i–or fails and yields K with probability $(1 - p_i)$. Thus the worst that can happen is that the project return equals the project cost. Furthermore, assume that all projects have a common expected payoff or gross return of $E(R_i)$ equal to M; this is an important assumption, whose role in the analysis we shall discuss in section 2.4. For now, however, note that, for all projects, the following holds:

$$E(R_i) = p_i R_i^s + (1 - p_i)K = M \qquad (2.1)$$

Entrepreneurs know both the probability of success and the value of the successful outcome associated with their project. Banks, on the other hand, are assumed to be ignorant of the probability of success, p_i, and the value if successful, R_s^i, of the ith borrower's project. We assume, however, that although banks are ignorant about the characteristics of any individual entrepreneur's project, they do know the distribution of the different types of project across the population of entrepreneurs and the value of the common expected gross return, M. Since, as we shall soon see, banks would prefer to lend to entrepreneurs with some types of project rather than others, but are unable to distinguish between them, they face a selection problem.

It will be easier to see what is going on if we consider a specific example. Let K be 100 and M be 120. Imagine that projects are of only two types, type 1 and type 2. Type 1 projects pay 130 if successful and have a success probability of 2/3 (which, as the reader may check using equation (2.1), yields $E(R) = M = 120$). Type 2 projects pay 140 if successful and have a success probability of 1/2 (again, the reader may check that the expected return is 120). Let there be n_1 entrepreneurs with type 1 projects and n_2 entrepreneurs with type 2 projects.

The standard debt contract

Assume, for simplicity, that an entrepreneur can borrow from only one bank, from which he must borrow K if he is to fund his project, and that all loans take the form of a *standard debt contract*. A standard debt contract is one in which the borrower pays the specified amount $(1 + r)K$ in the period after he has borrowed K if the project succeeds, or else pays the entire project return, in our case K, if the project fails. Banks, therefore, in our example, receive either K or $(1 + r)K$ once the project has been carried out, depending on whether it fails or succeeds, while borrowers receive either zero or any payoff in excess of $(1 + r)K$, that is $R_s^i - (1 + r)K$ in our example. The payment $(1 + r)K$ may be said to cover the repayment of the loan, or principal, of K, and interest on the principal, rK. The interest rate r, expressed as a per centage, is known as the *quoted loan rate* or the *interest rate charged on loans*.

Risk neutrality

We assume that entrepreneurs, banks and the suppliers of capital, who place funds on deposit at banks, are *risk neutral*. A risk neutral individual or firm is concerned only with the expected yield from an uncertain situation, and not in the possible payoffs and their associated probabilities which produce that expected yield. For example, a risk neutral person would be indifferent to a payoff of 50 dollars with certainty, or a gamble offering either a payoff of zero with a 50 per cent probability or 100 dollars with a 50 per cent probability, or another gamble offering either a payoff of zero with a 75 per cent probability and a 25 per cent probability of a payoff of 200 dollars, since in each case the expected or average value of the payoff is 50 dollars and that is all the individual is concerned about.

In some cases the assumption of risk neutrality is inappropriate; for example, in the analysis of insurance where individuals are concerned to avoid risk. However, assuming risk neutrality simplifies the analysis of the market for investment finance without preventing us from looking at the most important issues arising from asymmetric information. We can, in any case, imagine that banks invest in a sufficiently large number of projects, so that, although any individual project will yield the bank an uncertain return, it is able to spread its risks and estimate its average return per dollar loaned with certainty. In other words, by investing in a large number of projects, the bank avoids risk in terms of the aggregate return it expects to make. Banks can therefore offer to pay the suppliers of funds who place deposits with them a certain return also, so that they take no risks when placing their funds on deposit with a bank. Our assumption of risk neutrality, therefore, only really matters for the entrepreneur, who is unable to convert the uncertain payoff he will receive from his project into a certain payoff.

The demand for funds

Since the assumption of risk neutrality implies that an entrepreneur is concerned only with the expected payoff from investing in his project and all projects have the same expected return, it might seem that if one entrepreneur wished to obtain funds for his project, then so would all other entrepreneurs. This is not, however, the case since the standard debt contract introduces an important asymmetry in payoffs which makes only some, but not all, entrepreneurs wish to fund their projects at some values of the interest rate charged on loans. Equally, as we shall see later, the nature of the standard debt contract makes banks prefer to finance some types of project rather than others.

The entrepreneur's expected profit from borrowing to carry out his project, $E(\Pi_i)$, is given by:

$$E(\Pi_i) = p_i[R_i^s - (1 + r)K] \qquad (2.2)$$

which is just the probability of success times the amount received if successful under the terms of a standard debt contract with an interest rate charged on loans of r. An entrepreneur wishes to borrow from the bank to carry out his project as long as he expects to make a profit from doing so; that is, as long as:

$$E(\Pi_i) \geq 0 \qquad (2.3)$$

Constraint (2.3) is known as the *participation constraint*. It is so called because it must be satisfied before the agent to whom it applies is willing to participate in the market. When the constraint holds with equality, the entrepreneur is strictly indifferent to funding his project or not funding it, but we assume that in this case he does apply for funds. Another term sometimes used for the participation constraint is the *individual rationality constraint*, since it expresses the idea that a rational individual will only participate if he expects to make a gain (or at least break even on average) from doing so.

It is possible to see from equation (2.2) that the participation constraint (2.3) will be satisfied when R_i^s is greater than or equal to $(1 + r)K$. The higher the interest rate, r, on a loan, the higher will R_i^s need to be before an entrepreneur finds it worth borrowing to carry out his project. However, it follows from equation 2.1 that the larger is R_i^s, the smaller will be p_i. Thus we see that, under the standard debt contract, the entrepreneurs who choose to try to borrow to carry out their projects at any interest rate are those whose projects have a sufficiently large payoff if successful but an associated small chance of success. The interest rate, r, therefore, may be said to act as a *selection mechanism*, since it induces entrepreneurs to select themselves into two different categories: those who choose to apply for loans, and those who do not.

The selection mechanism may be shown very clearly if we consider our specific example. The participation constraint will be satisfied for all entrepreneurs for r less than or equal to 30 per cent, which can easily be seen by substituting the values for R_s^i and K in equation (2.2) for r less than or equal to 30 per cent. Therefore, the market demand for loans for interest rates up to 30 per cent will equal $100 (n_1 + n_2)$, since all entrepreneurs seek funding and each seeks to borrow the amount 100. Once the interest rate rises above 30 per cent, however, the participation constraint is no longer satisfied for entrepreneurs with type 1 projects, who cease to apply for loans. Entrepreneurs with type 2 projects, however, continue to apply for loans as long as the interest rate is no greater than 40 per cent. The market demand for loans for interest rates between 30 and 40 per cent is, therefore, $100n_2$, and for rates above 40 per cent the demand is zero. The market demand curve for loans, D_L, is shown in Figure 2.1, which plots the quoted loan rate, r, along the vertical axis and the market demand for loans, Q, along the horizontal axis.

Competition in banking

Banks are assumed to operate in a perfectly competitive industry, by which we mean that any supernormal profits made by banks are eliminated by competition.

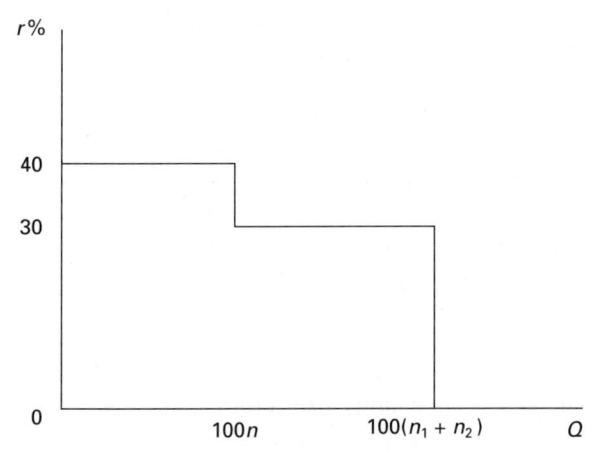

Figure 2.1 The demand for loans

Thus, if banks were making supernormal profits they would compete against one another to try to attract more deposits and make more loans. They would do this by offering higher deposit rates to attract more deposits, or by charging lower loan rates to attract more borrowers, or by a combination of these options. This process would continue until banks were making only normal profits. For simplicity, we assume that banks face no operating costs and that the level of normal profits is zero. Hence, in competitive equilibrium, banks must be making no profits, which means that the *deposit rate*, or the rate of interest they pay depositors, d, must equal the average rate of interest they receive from lending to entrepreneurs, ρ, which we shall call the *effective interest rate*. The constraint that in equilibrium:

$$d = \rho \tag{2.4}$$

is often known as the *competition or zero profit constraint*, since we impose it because of the nature of competition in the banking industry and use the simplifying assumption that normal profits are zero. We shall see below that there may be cases where zero profits are being made, but where competition and the search for positive profits would dictate that a bank, or principal, would change its actions in an attempt to gain supernormal profits. In other words, the zero profit constraint may be satisfied in circumstances where competition will force changes on the economy. Thus, satisfying the zero profit constraint is a necessary but not sufficient condition for the market to be in equilibrium.

The rate of interest banks charge and the return they receive

The effective rate of interest, or the average rate of interest banks receive from lending to entrepreneurs, ρ, is not equal to the interest rate which they charge borrowers, or the quoted loan rate, r. This is easily seen by noting that the bank's expected gross return from lending to fund a project of type i under our assumptions about project returns is given by:

$$E_b(\Pi_i) = p_i(1 + r)K + (1 - p_i)K = K(1 + rp_i) \qquad (2.5)$$

which is just the probability of success times the repayment if successful plus the probability of failure times the repayment if unsuccessful, which we have assumed to be K.[1] Hence the expected per centage return from funding a project of type i, which we denote by ρ_i, is given by:

$$\rho_i = rp_i \qquad (2.6)[2]$$

The higher the success probability of a project, the higher the expected per centage return the bank expects to make from it for a given interest rate. Therefore, at any interest rate, a bank prefers to lend to fund projects with a higher success probability than to fund projects with a lower success probability. On the other hand, for projects with a given success probability, the expected per centage return to banks increases as the interest rate charged on loans increases.

We saw above that the interest rate charged on loans acts as a selection mechanism, such that increases in it cause entrepreneurs with the higher probability of success to withdraw from the market. Thus an increase in the interest rate may produce two conflicting effects from the point of view of a bank. One effect is to raise the expected returns from those projects for which entrepreneurs continue to apply for funds; this effect is clearly beneficial for the bank. The other effect is to cause some entrepreneurs with projects with high chances of success to choose to withdraw from the market for funds, since $(1 + r)K$ becomes greater than R_i^s as the interest rate rises. This second effect gives rise to the phenomenon of *adverse selection*, which is the phenomenon that as the bank raises the interest rate it selects as applicants for loans only the entrepreneurs with projects that it views as the worst projects, since those with better projects (from the bank's point of view) withdraw from the market. If the second effect outweighs the first, then the bank reduces the average rate of return it makes on loans as it raises the interest rate it charges.

To see how the average rate of return for a bank can fall as the quoted loan rate is increased, consider once more the specific example given above with type 1 and type 2 projects. The average rate of return the bank makes on loans made at various quoted interest rates is easily calculated, as follows:

$$\begin{aligned} \rho &= r(n_1 p_1 + n_2 p_2) / (n_1 + n_2) & 0 \leq r \leq 30\% \\ \rho &= rp_2 & 30\% < r \leq 40\% \end{aligned} \qquad (2.7)$$

Thus, up to a quoted interest rate of 30 per cent, the average effective rate of interest, ρ, is just r times the weighted average probability of success, on the grounds that the bank lends to a sufficiently large number of entrepreneurs so that its loan portfolio contains the same proportions of entrepreneurs with type 1 and type 2 projects as the underlying population of entrepreneurs and all entrepreneurs apply for loans at quoted rates of interest up to 30 per cent. For rates of interest between 30 and 40 per cent, the bank lends only to fund type 2 projects, and the effective rate of interest is then given by rp_2, and for quoted rates of interest beyond 40 per cent all entrepreneurs cease to apply for funds.

At an interest rate of 30 per cent, the expected return to a bank from loans to fund type 1 projects is 20 per cent; that is the banks have a 2/3 probability of a payoff of 30 per cent (if the project is successful and the entrepreneur can pay off his debt) and a 1/3 probability of receiving a zero rate of return.[3] The percentage rate of return to be expected from a type 2 project at a 30 per cent interest rate can be calculated similarly as 15 per cent. The average effective rate of return, ρ, therefore, at a quoted loan rate of 30 per cent, lies somewhere between 15 and 20 per cent depending on the numbers n_1 and n_2 of entrepreneurs with projects of each type in the population. For simplicity, assume from now on that there are equal numbers of entrepreneurs with type 1 and type 2 projects, so that ρ is 17.5 per cent when r is 30 per cent.

As the interest rate is increased above 30 per cent only entrepreneurs with type 2 projects continue to apply for loans. Hence ρ falls to 15 per cent as the interest rate r passes 30 per cent, and then proceeds to rise towards 20 per cent as r is increased to 40 per cent. The reduction in ρ as r goes above 30 per cent is the result of adverse selection as entrepreneurs with type 1 projects, which yield the banks a higher expected return at an interest rate of 30 per cent, withdraw from the market.

The relationship between ρ and r given by equation (2.7) is shown in Figure 2.2, which plots ρ on the vertical axis and r on the horizontal.

The supply of deposits and loans

The analysis is greatly facilitated if we make the assumption that banks can acquire any amount of deposits they require as long as they provide a rate of

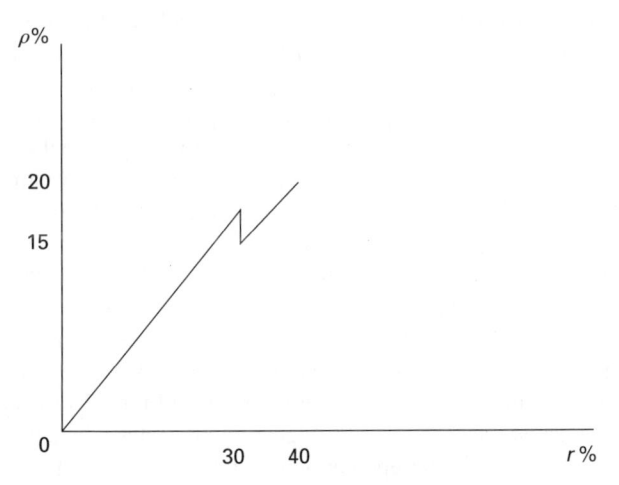

Figure 2.2   The relationship between ρ and r

14 Investment Finance

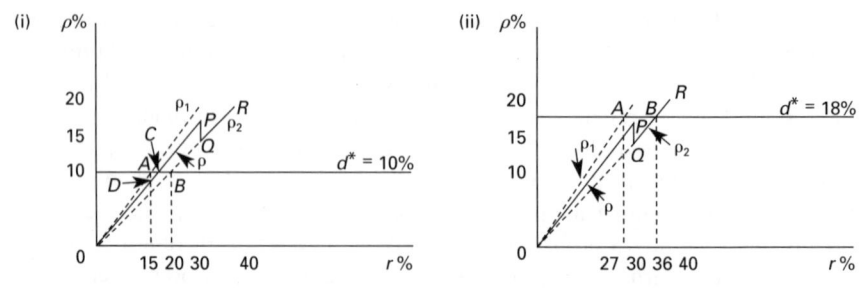

Figure 2.3 The credit market and the selection problem

return on deposits of d^*. Since we are assuming perfect competition in banking, this implies that banks will be willing to supply loans to entrepreneurs as long as the effective interest rate on loans, ρ, is equal to d^*.

Equilibrium in the loan market

We are now in a position to examine the operation of the credit market in our example, which we do with the aid of Figure 2.3. The figure is in two parts, each of which deals with different possible outcomes. Each part plots ρ on the vertical axis against r on the horizontal axis and shows three upward-sloping lines. The steepest of these three lines plots the relationship between ρ_1 and r; that is, it shows the effective return produced for a bank from lending to an entrepreneur with a type 1 project at various interest rates. Using equation (2.6) and the values for our example this is the ray through the origin, ρ_1 equals $(2/3)r$; it does not continue for values of r in excess of 30 per cent by virtue of the participation constraint. The least steep of these three lines shows the relationship between ρ_2 and r, that is ρ_2 equals $(1/2)r$, which does not continue beyond 40 per cent. Both these rays show an effective rate of return of 20 per cent at the respective quoted interest rates at which entrepreneurs choose to drop out of the market for loans. The reason for this is obvious: when the quoted interest becomes so high that the entrepreneurs' participation constraints hold with equality, so that their expected returns from participation become zero, then all the expected returns from the projects must be going to the banks. Since both types of project offer expected gross returns of 120 for an investment of 100, they both offer the banks an expected rate of return of 20 per cent when the quoted loan rates are so high as to offer zero expected returns to entrepreneurs.

The flash-shaped discontinuous relationship, $OPQR$, shows the expected effective rate of return produced at various interest rates for a bank lending to a random mixed sample of entrepreneurs with type 1 and type 2 projects. It is clear that up to a quoted loan rate of 30 per cent this relationship lies between the two rays representing ρ_1 and ρ_2, since entrepreneurs with type 1 and type 2 projects apply for loans at these interest rates, and that for higher interest rates it becomes coincident with the ρ_2 ray, since only entrepreneurs with type 2 projects apply for loans at these interest rates. The horizontal line at ρ equals d^* simply represents the required rate of return that banks must pay to depositors to attract funds, and which is, therefore, the effective rate of return that must be made on loans in equilibrium.

Consider, first of all, Figure 2.3 (i). Imagine initially that banks can distinguish between entrepreneurs who have type 1 and those who have type 2 projects. In this case, to achieve an effective return of d^*, which we assume to be 10 per cent, banks would charge 15 per cent to entrepreneurs with type 1 projects and 20 per cent to entrepreneurs with type 2 projects. This is shown by the intersections at A and B, respectively, of the horizontal line and the rays from the origin representing ρ_1 and ρ_2.

However, when banks are unable to observe the type of project with which entrepreneurs are associated, they cannot offer these two different interest rates. If they did so, entrepreneurs with type 2 projects would pretend that they had type 1 projects and take out loans at the lower interest rate of 15 per cent. The effective rate of return made by banks would then be given by point D on the $\rho - r$ relationship, which the reader, using equation (2.7), may verify, produces a rate of return to the banks of 8.75 per cent. Hence, banks lending to a mixed portfolio of borrowers at a quoted rate of 15 per cent would not be able to pay their depositors the required rate of 10 per cent and would go out of business. Instead, when banks are unable to observe the type of project available to an entrepreneur they must charge the quoted interest rate given by point C at the intersection of the horizontal line and the $\rho - r$ relationship. The reader may verify, using equation (2.7), that this interest rate is 17.14 per cent when there are equal numbers of entrepreneurs with type 1 and type 2 projects, which produces an effective rate of return of 10 per cent when the banks lend to a group of borrowers half of whom have type 1 projects and half have type 2.

It is easy to see that the effect of the asymmetry of information in this case leads to entrepreneurs with type 1 projects paying a higher quoted interest rate, and hence a higher effective rate, than they would in a full information world, while entrepreneurs with type 2 projects pay less than they would in a full information world. Thus the asymmetry of information imposes costs on entrepreneurs with type 1 projects, and provides benefits for those with type 2 projects who are, effectively, subsidised by those with type 1 projects. Clearly, entrepreneurs with type 1 projects will be unhappy with this outcome. We shall see in section 2.3 that entrepreneurs with type 1 projects will be able to do something about this situation and that the credit market will be replaced by an equity market, but before examining such arguments let us look at Figure 2.3 (ii).

This part of the figure differs importantly from part Figure 2.3 (i) by making the effective rate of return that banks must earn equal to 18 per cent. The significance of this is that the horizontal line representing d^* now lies above, rather than below, the kinked part of the average $\rho - r$ relationship.[4] In a world of full information the only impact of this is that the respective quoted loan rates rise for entrepreneurs, whether they have type 1 or type 2 projects, as shown by the intersections at A and B; the respective rates are 27 and 36 per cent, as the reader ought to be able to verify.

16   *Investment Finance*

In the asymmetric information case, the fact that the horizontal line lies above the kink in the average $\rho - r$ relationship leads to a significantly different result than for Figure 2.3 (i). There is now no single quoted interest rate that will attract both entrepreneurs with type 1 and entrepreneurs with type 2 projects to apply for loans, and that will yield an effective rate of return of 18 per cent. Hence, to earn a return of 18 per cent, banks must charge a quoted loan rate of 36 per cent, at which rate only entrepreneurs with type 2 projects apply for loans. There is no point such as *C* as there was in Figure 2.3 (i); instead, the relevant point is point *B*. Entrepreneurs with type 1 projects find their participation constraints are not satisfied at the interest rate of 36 per cent, which they find prohibitive, so they drop out of the market for credit.

The result in Figure 2.3 (ii) is worse than that in Figure 2.3 (i). In Figure 2.3 (i), point *C* represents entrepreneurs with type 1 projects subsidising those with type 2 projects; this result is disliked by those with type 1 projects but, is at least efficient in the sense that all projects, all of which are capable of offering a gross rate of return of 20 per cent which is greater than the rate required to attract funds, are funded. In Figure 2.3 (ii), however, at point *B* only type 2 projects receive funding, while type 1 projects, which are capable of offering a return greater than that needed to attract funds, do not get funded. This outcome is clearly inefficient, but cannot be avoided in the credit market in this case. Charging a low enough quoted loan rate to attract entrepreneurs with type 1 projects to seek funding would mean also attracting those with type 2 projects applying for loans at that rate and the banks on average being unable to cover the cost of attracting deposits.[5] Notice also that in this case entrepreneurs with type 1 projects suffer as a result of the asymmetric information compared to the outcome under full information, but now there is no offsetting gain made by entrepreneurs with type 2 projects. The market problems caused by asymmetric information in this case lead to what might be called a *dissipative externality*, which occurs when the information problems lead to costs being imposed on some people without anyone else gaining as a result.

The above results, although simple to follow and illustrative of the types of problem possible once asymmetric information is introduced, only go part of the way towards a proper analysis. These results are relevant if we assume that investment finance must be provided by borrowing under a standard debt contract. But, as the next section shows, such a contract would not be the best way to do business in the presence of the selection problem. Instead, market forces will lead to the development of equity finance, which will enable entrepreneurs with type 1 projects to avoid either subsidising those with type 2 projects as in Figure 2.3 (i), or to avoid being forced out of the market for funds as in Figure 2.3 (ii).

2.3   The selection problem and equity finance

The problem with the use of credit to finance investment projects in the face of the selection problem is that it is not always possible to set an interest rate to

The Selection Problem

enable the holders of projects to pay the required expected return necessary to attract funds. For example, in Figure 2.3 (ii), funds would be forthcoming to fund all projects if the interest rate paid to depositors was 18 per cent. All projects are capable of offering a return greater than this but cannot do so via the credit market; type 1 projects offer an expected return of 18 per cent to the banks when charged 27 per cent, but at a 27 per cent quoted interest rate the banks also find entrepreneurs with type 2 projects applying for loans and the average return to them is thus reduced to below 18 per cent. Hence banks charge a rate of interest of 36 per cent and type 1 projects go unfunded.

This problem can be overcome by the use of equity finance instead of credit finance. Indeed, equity finance would also be introduced instead of credit finance in the case of Figure 2.3 (i), as we shall see below. The intuition for the usefulness of equity finance is that, since all projects have the same expected return, they will all be valued equally by shareholders. Thus the use of equity finance removes the problem of selecting between different projects by making them all equally attractive.

Let us illustrate this point by considering how equity finance solves the problem posed in Figure 2.3 (ii). Starting from the credit market at point B in the figure, consider the response of entrepreneurs with type 1 projects who find the interest rate of 36 per cent to be too high to satisfy their participation constraint. Although forced out of the credit market, these entrepreneurs can try to attract funds by selling shares in the return to their project.

Remember that type 1 projects have a 2/3 chance of success and a successful payoff of 130. Each entrepreneur with a type 1 project can offer for sale a number of equities or shares, promising that the proceeds of the share sale will be invested in his project and that each share entitles its owner to 1 per cent of the gross return of the project. Let the price that the owners of funds would be willing to pay for such a share be V. The value for V_i will be given by the following formula:

$$V_i = [p_i R_i^s + (1 - p_i)K] / 100(1 + d^*) \qquad (2.8)$$

The right hand side of equation (2.8) is just the expected return of the project divided by 100, since each share is for 1 per cent of the firm, and then divided further by $(1 + d^*)$, since the suppliers of funds require an expected return of d^* on the funds they invest. It may, therefore, be rewritten as:

$$V_i = E(R)/100(1 + d^*) \qquad (2.9)$$

Since both types of project yield the same expected gross return, $E(R)$, in our example, then we may drop the i subscript from V_i. For both type 1 and type 2 projects, equation (2.9) yields V equals $(1.2)/(1.18)$, or 1.0169 for our example with d^* equal to 18 per cent. Hence the value of V, which the suppliers of funds would be willing to pay in order to achieve an expected return on shares of d^* is greater than 1.

18   *Investment Finance*

Notice that V is calculated under the assumption that the suppliers of funds, that is, the purchasers of shares, are risk neutral. If the reader thinks that this is an unreasonable assumption and that risk averse individuals would prefer to invest via a bank than to buy shares which seem riskier, then it is possible to imagine that the shares in our analysis are bought by unit trust companies or mutual funds. Such companies invest in the projects of so many different entrepreneurs that they do, in fact, achieve the required average rate of return d^* with certainty. Thus investment via unit trusts could enable investors to spread their risks just as effectively under equity finance as they could using banks under credit finance.

Entrepreneurs with type 1 projects would be willing to obtain equity finance for their projects as long as:

$$V \geq 1 \tag{2.10}$$

Constraint (2.10) holds, since as long as V is greater than or equal to 1, entrepreneurs can fund their projects while retaining an equity stake and a positive expected return for themselves. Constraint (2.10) is the participation constraint for entrepreneurs seeking equity finance. The participation constraint is clearly satisfied for entrepreneurs with type 1 projects, who would offer to sell shares instead of trying to acquire credit.

Starting from point B in Figure 2.3 (ii), the suppliers of funds would, in fact, be indifferent between supplying funds to banks, which lend at 36 per cent and offer an average return of 18 per cent, or buying shares that yield an expected return of 18 per cent. Entrepreneurs unwilling to obtain funds from the credit market could, therefore, attract funds by selling shares at a price equal to 1.0169, offering shareholders an expected return of 18 per cent. One response for entrepreneurs with type 2 projects would be for them also to offer to sell shares at that price so that all projects would eventually be funded via the share market. Alternatively, entrepreneurs with type 2 projects could continue to borrow at a quoted loan rate of 36 per cent, such that in equilibrium type 1 projects could be share financed and type 2 projects could be credit financed as long as the expected rate of return to the suppliers of funds was equal to 18 per cent whether they funded type 1 or type 2 projects. Entrepreneurs with type 2 projects in this mixed equilibrium would be indifferent to acquiring credit or equity finance. Entrepreneurs with type 1 projects would seek only equity finance.

In equilibrium, in our example, a rate of return on shares of 18 per cent would be sufficient to attract funding for all projects. The equilibrium share price of 1.0169 would mean that each entrepreneur will retain an equity stake in his project, since he needs to sell fewer than 100 shares to fund it. In fact, to fund his project, an entrepreneur would need to sell 100/1.0169 shares, that is 98.33 shares, which would leave the entrepreneur an equity stake of (100 minus 98.33) per cent, that is 1.66 per cent, in his project. The entrepreneur would, therefore, make an expected gain of 1.66 per cent of the expected gross return of 120 of his project; that is, an expected return of 2. The reader may verify, by using equation (2.2), that this is exactly equal to the expected return of an entrepreneur with a type 2 project borrowing at a quoted loan rate of 36 per cent. The remaining 118 of the expected gross return of the project is the expected gross return to the suppliers of funds, either shareholders or creditors. Thus it would be possible either for all projects to be equity financed or for all type 1 projects to be equity financed and some or all type 2 projects to be credit financed. If a mixed equilibrium of credit and equity finance occurred it would have no effect on expected returns compared to the equity finance case: in either case all projects would offer d^* equal to 18 per cent to the suppliers of funds and an expected return of 2 to entrepreneurs.[6]

The share market equilibrium (assuming the absence of a credit market) may be illustrated diagrammatically using Figure 2.4. The supply schedule for shares, S_S, is the kinked line shown. The shape may be explained by the fact that the participation constraint given in constraint (2.10) tells us that no shares will be offered for sale for a price of less than 1 (since then it would be impossible to fund the project while leaving the entrepreneur a stake in it). On the other hand, at a price greater than 1 all entrepreneurs would be willing to sell shares in their projects, since then they could fund their projects while keeping a potentially profitable stake for themselves, which is preferable to not funding their projects. Hence, for any price greater than 1, the supply of shares (or the demand for funds) is $100(n_1 + n_2)$, which is the amount required to fund the projects of all entrepreneurs.[7]

The demand schedule is the horizontal line given by V equals $E(R)/100$ $(1 + d^*)$, which is at a value of 1.0169 for our example. Equilibrium is shown at the intersection of the supply and demand schedules at point E.

In the equity market equilibrium of Figure 2.4 there is no longer an efficiency problem as in Figure 2.3 (ii), where projects capable of paying more than the market rate of return to the suppliers of funds go unfunded, since all projects are now funded. Thus the use of equity finance in our example solves the problems posed by asymmetric information for the credit market. The reason is simple: under equity finance all projects are equally good from the point of view of the suppliers of funds, since all yield the same expected return for the same price per share. With equity finance there is no problem of adverse selection and the market does not suffer from the asymmetry of information as it would under credit finance. It would also be possible for a mixed equilibrium to exist, in which type 2 projects were funded by credit, where they would offer the same return to the suppliers of funds as if they were equity financed, as we made clear above.[8]

The reader may wonder why it is that the equilibrium share price is above the minimum price of 1 at which entrepreneurs would be willing to sell. The reason is that, under our assumptions, if the share price was 1, buyers would demand more shares than would be made available and would compete among themselves to push up the price to the level at which supply and demand are equal. Formally, this idea may be expressed as saying that for any value of Q, the share price will be given by equation (2.9). Since it is competition which forces the economy to an equilibrium at which equation (2.9) determines the price, it is possible to call equation (2.9) the zero profit or competition constraint for the equity finance case.

Notice that entrepreneurs with type 1 projects, who gain funds under equity finance but are driven from the market by prohibitive interest rates under credit finance, clearly gain from the introduction of the equity market, and more suppliers of funds are able to earn the expected rate of return necessary to attract them to provide funds. Entrepreneurs with type 2 projects who obtained credit under credit finance are no worse off under equity finance, since they pay the same expected rate of return to the suppliers of funds in either case. The introduction of the equity market is, therefore, clearly a good thing. In other words, the movement from a credit market to an equity market (or a mixed equity/credit market) represents a *Pareto improvement*, which is a change which benefits some members of the economy and harms none.

Thus, the credit market position in Figure 2.3 (ii), in which not all projects are funded, would be destroyed by entrepreneurs without funding offering to sell shares. The return to society as a whole is increased by funding all projects under equity finance, since this allows projects with a rate of return greater than the equilibrium rate offered to depositors under credit finance to be funded.

The equity market equilibrium, once achieved, would not be destroyed by banks or entrepreneurs offering or seeking credit finance rather than equity finance. This is so because, from the equity market equilibrium, no entrepreneur could offer investors a higher rate of return without at the same time reducing his own expected return, and therefore none would seek credit finance once the equity market equilibrium, was achieved. On the other hand, starting from the equity market equilibrium, no supplier of funds could offer a credit arrangement which would increase the supplier's expected return without at the same time reducing the entrepreneur's expected return compared to the equity arrangement. Thus the equity market equilibrium would not be vulnerable to anyone trying to use credit to replace equity finance.

The equity market equilibrium is an example of a *Nash equilibrium*, which is a well-known concept in game theory. A Nash equilibrium occurs when no player (in our case the entrepreneurs and the suppliers of funds) in a game has an incentive to behave differently (or, in other words, to deviate from his strategy or choice of action), given the behaviour of the other players (or, in other words, given the strategies of the other players). A Nash equilibrium is, therefore, robust in the sense that once achieved no player has an incentive to move away from it (unless he perceives that he has market power and that others will be forced to react to his move, which we rule out by assuming that all individuals are small relative to the market as a whole). The concept of Nash equilibrium will be used frequently throughout this book.

It would be possible in a Nash equilibrium for mixed finance to occur, with some type 2 projects being funded by credit finance and type 1 projects and the remaining type 2 projects being equity financed. In such a mixed equilibrium the expected returns for the suppliers of funds, and the entrepreneurs, would be

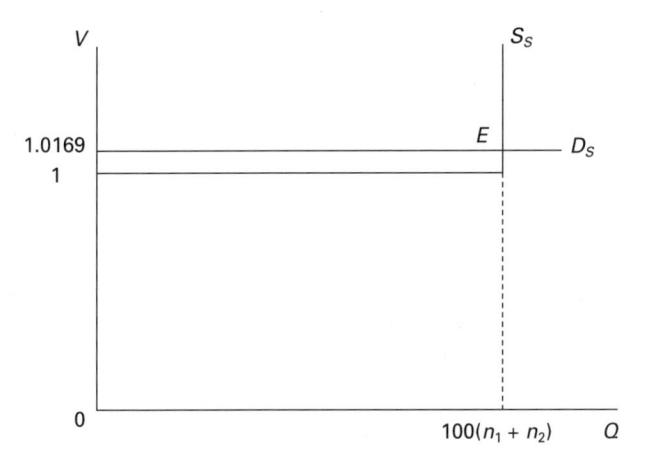

Figure 2.4 The equity market

identical to the returns offered in the equity market equilibrium, as we saw above. Since the credit market in Figure 2.3 (ii) would not be robust in the face of entrepreneurs offering shares rather than seeking credit, then the credit market clearly does not represent a Nash equilibrium.

Notice also that the equity market (or mixed market) equilibrium is *Pareto efficient*; that is, it would be impossible to move away from it without harming at least one of the two sets of entrepreneurs, those with type 1 or those with type 2 projects, or the suppliers of funds. This is clearly the case, since all projects are being funded and the maximum returns for society are being gained.[9]

The credit market in Figure 2.3 (i) was also Pareto efficient, since all projects were funded. The credit market in Figure 2.3 (i) would not, however, be a Nash equilibrium and there would be forces operating to introduce equity finance. These forces would be introduced by entrepreneurs with type 1 projects. This is so because, in the credit market, entrepreneurs with type 1 projects pay a higher expected return to the suppliers of funds than do entrepreneurs with type 2 projects. Therefore, entrepreneurs with type 1 projects can offer shares which offer investors a higher expected rate of return than that offered by the credit market, while still being themselves better off than under credit finance (where they were subsidising entrepreneurs with type 2 projects). Competition between the suppliers of funds will then push this share price up until the rate of return offered to shareholders is just the 10 per cent we assume necessary to attract funds; the share price will then be 1.2/1.1, or 1.091. In equilibrium, either all projects are equity financed, in which case their share prices are equal and offer the suppliers of funds an equal expected return of 10 per cent, or else some type 2 projects are credit financed, but at a quoted loan rate (20 per cent in our case) offering the suppliers of funds the same expected return as if they were equity financed at the same share price as the type 1 projects (and any type 2 projects which are equity financed). In either case, entrepreneurs with type 1 and type 2 projects all offer the suppliers of funds the same return in the new equilibrium.

22   Investment Finance

This rate of return is the same as the average return d^*, equal to 10 per cent in our example, which the suppliers of funds were being offered in the credit market.

Remember that in the credit market entrepreneurs with type 1 projects were paying more than entrepreneurs with type 2 projects. Hence, entrepreneurs with type 1 projects gain from the levelling off of returns offered to the suppliers of funds brought about by the introduction of an equity market, and entrepreneurs with type 2 projects lose, while the suppliers of funds are unaffected as they gain the same expected return in both cases.

The equity market diagram which matches the credit market of Figure 2.3 (ii) would look like Figure 2.4, the only difference being that the share price which yields an expected return of d^* to the suppliers of funds is now 1.2/1.1, or 1.091, since d^* takes the lower value of 10 per cent rather than 20 per cent as for Figure 2.4.

2.4   Credit rationing

The supply of deposits

In order to produce credit rationing as a response to the selection problem it is necessary to relax only our assumption that the suppliers of funds are willing to supply as many funds as are demanded as long as they receive an expected rate of return of d^*. Instead, we assume that the supply of deposits placed with banks increases as the rate of interest, d, paid by the banks to depositors increases. We also consider the more general case where we do not necessarily assume equal numbers of entrepreneurs with type 1 and type 2 projects. To ensure that the rationing we derive below is produced by the asymmetry of information and not by any shortage of funds, we assume that the supply of funds at a deposit rate of 20 per cent, equal to the rate that all projects in our example are capable of providing, would be great enough to fund all projects.

The supply of deposits schedule, S_D, is shown in Figure 2.5, which plots the deposit rate, d, on the vertical axis and the quantity of deposits, Q, on the horizontal axis.

The supply of loans

Given our new assumption about the supply of deposits, and maintaining all our other assumptions, it is possible to derive a supply of loans schedule showing the amount of loans that would be provided at various quoted loan rates.

The supply of loans schedule, S_L, may be derived using Figure 2.6. Figure 2.6 (i) presents the $\rho - r$ relationship given by equation (2.7) and Figure 2.6 (ii) repeats Figure 2.5. Figure 2.6 (iv) simply presents a 45° line to translate the interest rate, r, from Figure 2.6 (i) to Figure 2.6 (iii) which shows the supply of loans. It is a property of the 45° line that, when the scales are the same on both

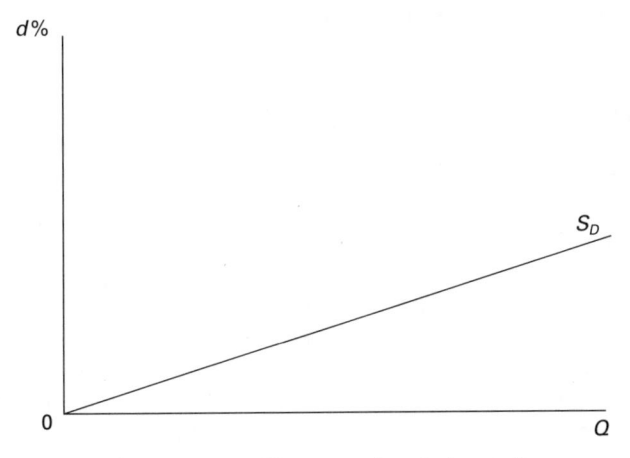

Figure 2.5 The supply of deposits

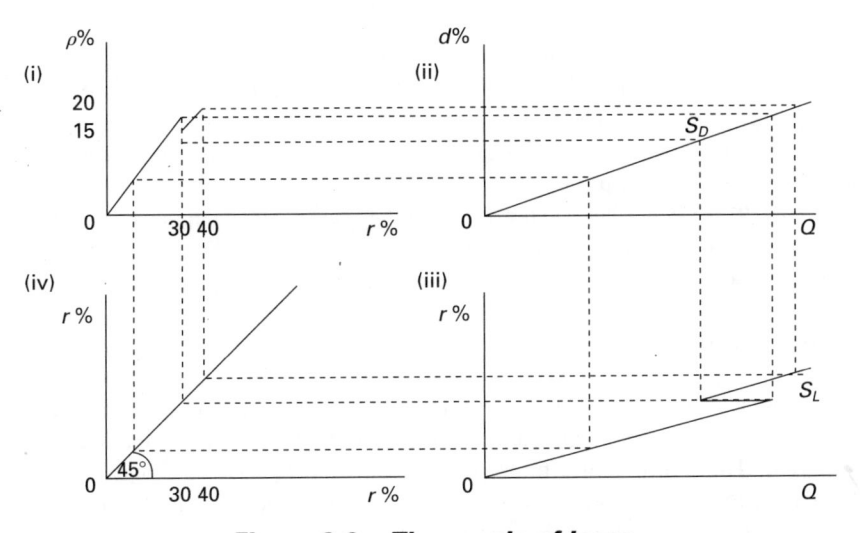

Figure 2.6 The supply of loans

axes, all points on it represent equal values for r on both vertical and horizontal axes. Thus, reading across from the 45° line to the intersection in Figure 2.6 (iii) with the line drawn down from Figure 2.6 (ii) correctly links r and Q values.

The relationship shown in Figure 2.6 (iii) is easily derived. Consider some interest rate, r, (say, 40 per cent) in Figure 2.6 (i) and read up to the average $\rho - r$ relationship to find the corresponding value of ρ. Reading across from the ρ value in Figure 2.6 (i) to the equal value for d in Figure 2.6 (ii), which we can do, since in equilibrium competition between banks will ensure that ρ will be equal to d. We can then read down to find the volume of deposits, Q, that could be attracted

24 Investment Finance

by the banking industry for the interest rate, r, with which we started. Since these deposits can be supplied as loans by the banks we can then plot the volume of loans, Q, against the interest rate, r, in Figure 2.6 (iii). The supply of loans schedule, S_L, in Figure 2.6 (iii) can be produced by reading down from Figure 2.6 (ii) to the intersection with the relevant interest rate, r, found by reading down from Figure 2.6 (i) to Figure 2.6 (iv) and then across to Figure 2.6 (iii).

Notice that the supply of loans so derived is discontinuous at an interest rate of 30 per cent. This is because of the discontinuity in the average $\rho - r$ relationship at that interest rate, which also shows up in the supply of loans function. The intuition is straightforward. The adverse selection, as entrepreneurs with type 1 projects withdraw from the market when the interest rate passes 30 per cent, causes the average rate of return to the banks to fall, and they can attract fewer deposits. As the interest rate rises beyond 30 per cent, the average rate of return to the banks rises once more and they begin to attract more deposits again. The supply curve, like the average $\rho - r$ relationship, stops at r equals 40 per cent, since at interest rates greater than 40 per cent no entrepreneurs seek funds and so no return could be made by lending at interest rates in excess of 40 per cent.

The reader ought to be able to check, by carefully drawing the diagram, that the slope of the S_L schedule is steeper for loan rates above 30 per cent than for rates below that value. The reason is that for interest rates below 30 per cent the pool of loan applicants is better from the bank's point of view. Hence a rise in r causes a bigger rise in ρ and helps to attract a bigger increase in deposits when r is below 30 per cent than when it is above that value. Since deposits are used to provide loans, it follows that the supply of loans rises faster as r rises when it is below 30 per cent than when it is above 30 per cent.

Equilibrium in the loan market

It is now possible to consider the equilibrium in the loan market by using Figure 2.7, which plots the supply and demand for loans schedules on the same diagram. The figure shows three possible outcomes, where the parts may be imagined to differ because the supply of deposits schedules and/or n_1 and n_2 may differ from one part to the next.

Figure 2.7(i) shows the case where the supply and demand curves intersect twice, at points A and B in the figure. Both points A and B represent points where supply equals demand, but B will unambiguously be the market equilibrium. This can easily be seen since the supply of funds is greater at B than at A, which means that the average rate of return to the banks, ρ, and the deposit rate, d, must be greater at B. Hence, any bank operating at A would be able to reduce its interest rate from r_A to r_B and attract both more entrepreneurs seeking loans and more depositors. Thus all banks are driven by competition to locate at point B rather than point A. In this case, the supply of funds is so plentiful that all projects are funded and the asymmetry of information has no harmful effects on efficiency.

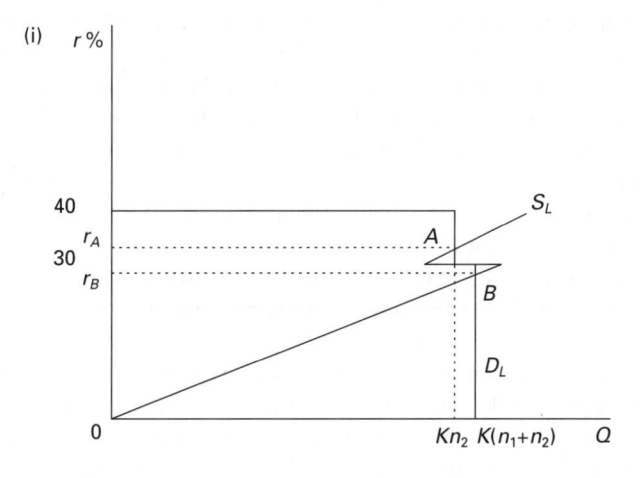

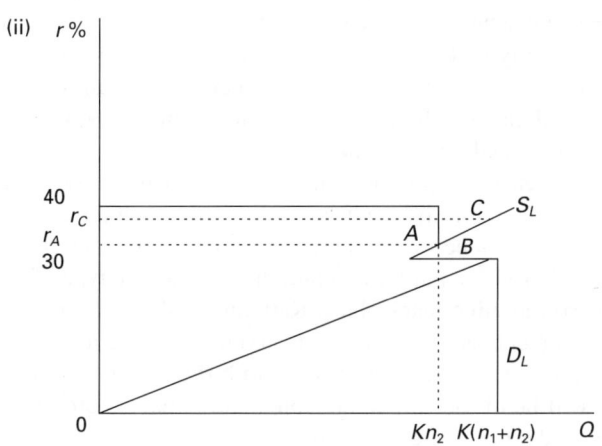

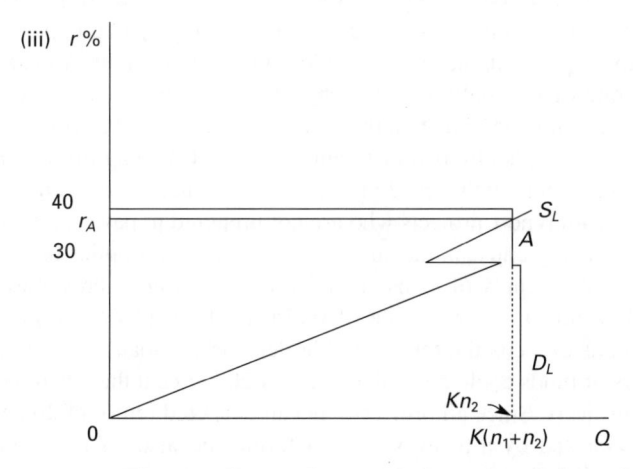

Figure 2.7   The credit market and the selection problem

Notice that all entrepreneurs would continue to demand loans if the interest rate was increased from r_B to 30 per cent, but competition between banks prevents the interest rate rising above r_B. This can be seen easily by imagining banks operating at an interest rate of 30 per cent, in which case they would be making supernormal profits, with p exceeding d. However, from such a position, any one bank would have the incentive to try to attract more entrepreneurs by reducing the interest rate on loans, other banks would follow suit and the process would continue until point B was reached. Alternatively, the reader may imagine that entry into the banking industry is unrestricted, so that if supernormal profits were being made by banks this would attract new entrants who would compete away those profits.

Figure 2.7(ii) shows a more problematic case. Here there is only one intersection of the supply and demand curves, at point A. At A, only entrepreneurs with type 2 projects apply for loans. However, point A will not be the market equilibrium, which will instead occur at point B, where the horizontal parts of the supply and demand schedules are coincident at the interest rate of 30 per cent. At point B the supply of loans exceeds the supply at A, therefore the average effective rate of return, p, and the associated deposit rate, must be higher at B than at A. Thus all banks will be driven by competition to reduce their interest rates from r_A to r_B (equals 30 per cent).

At B, all entrepreneurs apply for loans and the demand exceeds the supply of loans. There will, therefore be credit rationing at the market equilibrium at B. Banks would prefer to lend to entrepreneurs with type 1 projects rather than to those with type 2 projects, but they cannot distinguish one type of entrepreneur from another, so rationing takes place. Rationing will be determined by some arbitrary method (such as first-come-first-served or tossing a coin to determine who gets a loan). Since entrepreneurs need to borrow K to fund their projects, the rationing will be of the sort where some entrepreneurs are granted the full loan they seek and others are granted no loan at all.

Entrepreneurs with type 1 projects denied loans at B are not prepared to bid up the interest rate, but those with type 2 projects are prepared to pay an interest rate of up to 40 per cent and so would bid up the interest rate. A bank charging an interest rate of r_C would attract such borrowers and gain the same average return as at B, so it is possible that the market could be characterised not only by credit rationing but also by two different interest rates being offered, r_B and r_C, with all entrepreneurs with type 2 projects gaining funds and the rationing only hitting those with type 1 projects who are not prepared to pay r_C.[10]

Apart from being unusual, the above equilibrium also implies that the market is inefficient. The supply of loans curve shows that at an interest rate of 40 per cent, which yields an average rate of return to the banks of 20 per cent, the supply of loans exceeds the total possible demand for loans of $100(n_1 + n_2)$. If the suppliers of funds could be paid a return of 20 per cent they would be willing to fund all projects. Since all projects have an expected return of 20 per cent the rationing is caused by a form of market failure because of the informational asymmetry and not because of a lack of funds.

If lenders could distinguish between the two types of project for which funds were sought, they could charge borrowers with different types of project different interest rates, such that each type of project offered lenders just that expected return necessary to induce them to supply $100(n_1 + n_2)$ funds to the market. The trouble is, of course, that entrepreneurs with type 2 projects are not prepared to reveal that information and, if asked, would pretend to have type 1 projects. This problem would not, of course, arise if all individuals were perfectly moral, never told lies and always revealed the truth. Indeed, many of the problems caused by asymmetries of information could be avoided if individuals were always moral in this sense, but the problem is that individuals often have an incentive to hide information or to tell lies, and the market outcome reflects this.

Figure 2.7(iii) shows a more straightforward market equilibrium at point A. At this point only entrepreneurs with type 2 projects apply for loans and the market clears without rationing. Although more straightforward than the equilibrium in Figure 2.7(ii), the equilibrium at A is still inefficient. As we have drawn it, the supply of funds line shows that at an interest rate of 40 per cent, or a return to depositors of 20 per cent, all projects could be funded, so the problem is that the asymmetry of information is preventing the type 1 projects from being funded. As for Figure 2.7(ii) above, all projects could be funded if lenders could distinguish between the two types of project and could charge borrowers different interest rates according to their project type. Because entrepreneurs with type 1 projects are not willing to pay the high interest rate r_A, and banks are not prepared to lower the interest rate, since at the lower interest rate they attract entrepreneurs with both type 1 and type 2 projects and cannot distinguish between them, the credit market is unable to direct funds towards type 1 projects. Even if the supply of funds at a rate of return to depositors of 20 per cent was insufficient to fund all projects, the equilibrium in Figure 2.7(iii) would still be inefficient, with type 1 projects being capable of yielding more than the market return to the suppliers of funds remaining unfunded.

Comparisons with conventional markets

The equilibria illustrated in Figure 2.7 further indicate the potential for asymmetric information to produce results not usually found in the more conventional symmetric information analysis. The analysis differs from the conventional analysis in a couple of ways that are worth making explicit here. First, since the interest rate acts as a selection mechanism, it is clear that the quality of the loans made, from the banks' point of view, depends upon the interest rate. In general, in problems with asymmetric information, this result shows up as the dependence of the quality of the goods bought and sold on a market depending on the price charged. In such markets, the market price not only affects the quantity bought and sold, as in a conventional market, but it also affects the quality of goods being bought and sold. In some cases, such as in Figure 2.7(ii), the equilibrium does not equate quantity supplied and demanded and it is possible for there to be more than one price charged in equilibrium.

28   *Investment Finance*

Second, it is important to note that in conventional analysis it is usual to consider supply and demand independently and then to put them together in a supply and demand diagram to consider the equilibrium. We followed this traditional path as closely as possible in the above analysis, but it is worth noting that our derivation of the supply of funds line depended crucially upon the average $\rho - r$ relationship. This relationship depends upon the number of entrepreneurs with either type 1 or type 2 projects, and on the nature of the return distributions for those projects. Changing the nature of the projects or the relative proportions of entrepreneurs with type 1 or type 2 projects would therefore cause not only the demand curve to shift but would also affect the supply curve. Thus supply and demand schedules are not independent of one another as in the usual analysis. This is not to say that all changes to one schedule also affect the other one; for example, a shift in the supply of deposits schedule, S_D, would shift the supply of loans schedule, S_L, without shifting the demand for loans schedule, D_L.

As for the simpler analysis in section 2.1 above, our analysis of the credit market under the selection problem has allowed us to show quite clearly how asymmetric information can produce results not found under the conventional symmetric information assumption. The analysis, however, has assumed that entrepreneurs could only acquire funds by borrowing under a standard debt contract. In fact, as we saw for the simpler analysis in sections 2.2 and 2.3, such a contract would not be the best way to do business in the presence of the selection problem.

Equity finance

Equity finance can again be used to overcome the inefficiencies present under debt finance. The equity market equilibrium in this case may be illustrated using Figures 2.8 and 2.9.

Figure 2.8(i) shows the supply of funds available, Q, as a function of the expected return on those funds, d, which should no longer be interpreted as the deposit rate, although in every other way the supply of funds line is identical to that drawn in Figure 2.5 above. The greater the expected return, the greater is the supply of funds. Figure 2.8(ii) shows the relationship between the share price, V, and the expected rate of return on funds used to buy shares, d, given by equation (2.9) with d^* replaced by d. The maximum value of V is shown as 1.2, since at this value the purchasers of shares would be making a zero expected percentage return; that is, d would equal zero and we assume no funds would be supplied at this rate.

Figure 2.8(i) shows that if shares offered a rate of return of d_1, for example, then Q_1 funds would be made available for investment via the share market, while reading across to Figure 2.8(ii) shows that this return would be offered if the share price was V_1. Hence Q_1 would be made available on the share market if the share price was V_1, which produces the point Q_1, V_1 on the demand for shares schedule, D_S, in Figure 2.8(iv). This point on the demand schedule may be found by reading down from V_1 in Figure 2.8(ii) to the 45° line in Figure 2.8(iii) and

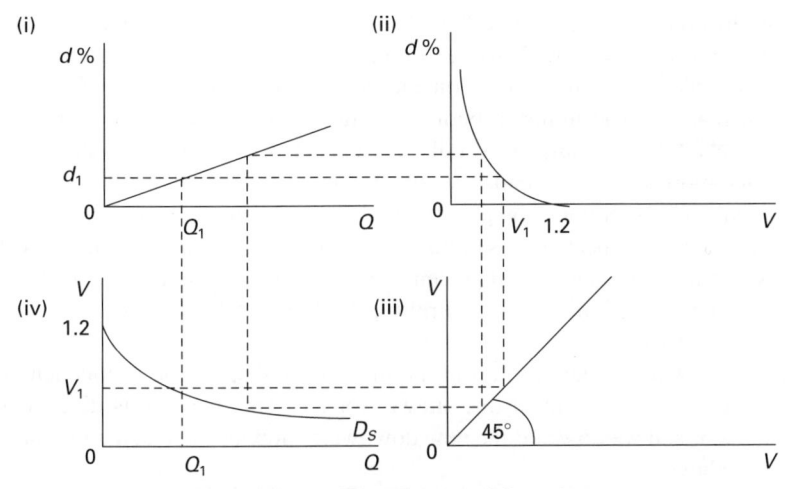

Figure 2.8 The demand for shares

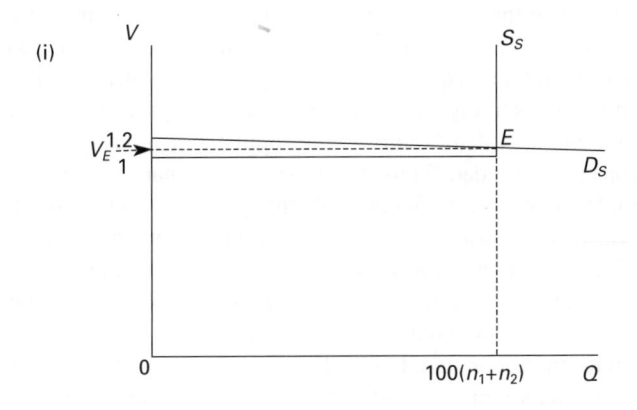

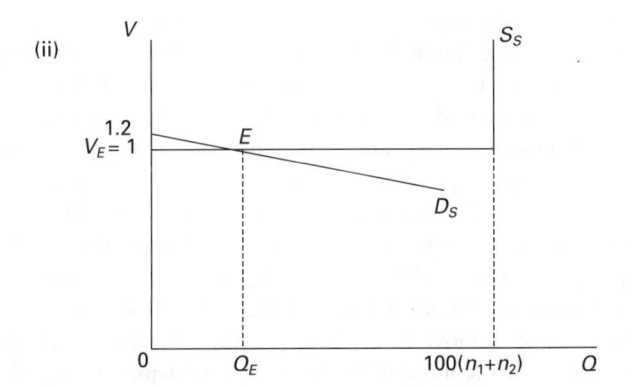

Figure 2.9 The equity market

across to Figure 2.8(iv), where the intersection with the line found by reading down from V_1 in Figure 2.8(i) to Figure 2.8(iv) produces the point on the demand schedule. Other points on the demand schedule may be found in a similar manner. The intuition behind the downward slope of the demand for shares schedule is simple: as V falls, the expected return from holding shares rises and more holders of funds are attracted into the share market.

Given our assumptions, the payoff per share will either be 1 (that is $K/100$) or $R_i^s/100$, with an expected gross return of 1.2 (that is, $E(R)/100$). The expected gross return yields an expected percentage return of d (given the price of V) and, given our assumption of risk neutrality, the buyer values shares at V equals $E(R)/100(1 + d)$.

Figure 2.9 shows equilibrium in the market for shares. The supply schedule for shares, S_S, is the kinked line shown, for the same reasons as discussed in section 2.3, and we now add the new downward-sloping share demand schedule derived above.

The figure shows two possible types of equilibria.[11] Figure 2.9(i) shows the case where, as in our analysis of the credit market using Figure 2.7, the suppliers of funds are willing to supply enough funds to fund all projects at a rate of return to them of less than the expected rate of return available on projects. Hence the equilibrium share price will be greater than 1 and all projects will be funded.

In the equity market equilibrium in Figure 2.9(i) there is no longer an efficiency problem as in Figure 2.7(ii) and (iii), where projects capable of paying more than the market rate of return to the suppliers of funds go unfunded, since now all projects are funded. Thus the use of equity finance in our more complicated example solves the efficiency problems posed by asymmetric information for the credit market, just as it solved the problem for the simpler example of sections 2.2 and 2.3. Just as for the simpler example, with equity finance there is no problem of adverse selection and the market does not suffer from the asymmetry of information as it would under credit finance.

Notice that since, compared with Figure 2.7(ii) and (iii), more funds are attracted to the market under equity finance than under credit finance, the equilibrium return to the suppliers of funds must be greater under equity finance than under credit finance. The suppliers of funds are therefore clearly better off under equity finance than under credit finance. Entrepreneurs with type 1 projects who gained funds under equity finance but were driven from the market by prohibitive interest rates under credit finance as in Figure 2.7(iii), or who were victims of rationing in Figure 2.7(ii), are also better off under the equity market. Entrepreneurs with type 2 projects, or those with type 1 projects in Figure 2.7(ii) who obtained credit under credit finance, are worse off under equity finance: this follows because the expected return they must now offer to shareholders exceeds that which they used to offer to depositors. Nevertheless, the equity market equilibrium is Pareto efficient, since, as for the simpler example, it would be impossible to move away from it without harming at least one of the two sets of entrepreneurs, those with type 1 or those with type 2 projects, or the suppliers of funds.

The Selection Problem 31

Thus the credit equilibria in Figure 2.7(ii) and (iii), in which not all projects are funded, would be destroyed by entrepreneurs without funding offering a higher return to investors who buy shares than to those who place their assets in banks; thus funds would be attracted away from the credit market and towards the equity market. The return to society as a whole is increased by funding all projects under equity finance, since this allows previously unfunded projects with a rate of return greater than the equilibrium rate offered to depositors under credit finance to be funded.

The equity market equilibrium, once achieved, would represent a Nash equilibrium exactly as for the simpler example in the earlier sections. Since the credit market outcomes in Figure 2.7(ii) and (iii) would not be robust in the face of entrepreneurs offering shares rather than seeking credit, then they were clearly not Nash equilibria. The reader should be able to argue that a mixed market with the use of both equity and credit could also represent a Nash equilibrium, and it should be clear that in such an equilibrium the returns to entrepreneurs or the suppliers of funds will not be affected by whether a project is financed by credit or by equity provided that entrepreneurs choose the form of contract which maximises their expected returns.

The equilibrium in Figure 2.7(i) was efficient, since all projects were funded. Even so, it would not be a Nash equilibrium and there would be forces operating to introduce equity finance. These forces would be introduced by entrepreneurs with type 1 projects. The reasons for this result parallel those presented in section 2.3 above in relation to Figure 2.3(i). In the credit market outcome, entrepreneurs with type 1 projects pay a higher expected return to the suppliers of funds than do entrepreneurs with type 2 projects. Therefore entrepreneurs with type 1 projects can offer shares that offer investors a higher expected rate of return than that offered by the credit market, while still being themselves better off than under credit finance (where they were subsidising entrepreneurs with type 2 projects). Hence, in equilibrium, either all projects are equity financed, in which case their share prices are equal and offer the suppliers of funds an equal expected return, or some type 2 projects are credit financed, but at a rate offering the suppliers of funds the same expected return as if they were equity financed at the same share price as the type 1 and the remaining type 2 projects that are equity financed. In either case, entrepreneurs with type 1 and type 2 projects offer suppliers of funds the same return in the new equilibrium. This rate of return is the same as the average return the suppliers of funds were being offered in the credit market outcome (since it attracts exactly the same amount of funds from the suppliers of funds), although, in the credit market, entrepreneurs with type 1 projects were paying more than entrepreneurs with type 2 projects. Hence entrepreneurs with type 1 projects gain and entrepreneurs with type 2 projects lose, while the suppliers of funds are unaffected because they gain the same expected return in either case, which is just the return necessary for them to supply $100(n_1 + n_2)$ funds to the market.

For completeness, Figure 2.9(ii) shows the case where entrepreneurs' projects are not capable of offering a sufficiently high rate of return to attract enough funds to fund all projects, even in the absence of informational problems. Equilibrium is shown at the intersection of the supply and demand schedules at point E. In this case, there will be an excess supply of shares at an equilibrium price of 1, so that not all projects will be able to acquire funding.[12] Despite the rationing, the equilibrium is efficient. The suppliers of funds who are willing to accept a return less than or equal to the expected return offered by projects will invest in projects, and those who would require a higher rate do not do so. The rationing of funds to projects would be decided arbitrarily, since all projects offer equal expected returns and nothing may be gained by directing funds to one type of project rather than another.

2.5 Discussion

The analysis above has shown how the selection problem can lead to adverse selection and inefficiencies in the credit market. We have also seen how the use of equity finance can solve these problems. The analysis has been based around a very simple case of the selection problem, however, and there are a number of points worth making to extend the analysis or relax the special assumptions we have made.

One point to note is that the standard debt contract we considered was very simple. For example, we did not allow entrepreneurs to invest any of their own wealth in their projects, or to offer their own wealth as collateral against their projects failing. The use of collateral or their own wealth by entrepreneurs might, in fact, allow banks to better select between projects – those entrepreneurs with safer projects being typically more willing to invest their own wealth or use it as collateral with a bank. Banks may therefore adjust the contract terms according to the entrepreneur's willingness to use his own wealth and so avoid some of the problems stemming from the inability to distinguish between entrepreneurs under the standard debt contract. Perhaps this mechanism partly explains the often expressed criticism of banks that they are apparently more willing to lend to borrowers who have more of their own wealth to invest in their projects. The use of their own wealth allows entrepreneurs to send a *signal* to banks about their project characteristics. *Signalling* is an important issue in the economics of asymmetric information, but we shall not consider it further here, since in our examples the use of equity finance removes the need for signalling, by removing the selection problem. We shall, however, have more to say on signalling in Chapter 8 below.

Another point to note is that the assumption that all projects have the same expected return is important for the occurrence of adverse selection and credit rationing. Intuitively, if projects differ in expected returns then some are, in terms of expected returns, better than others. Raising the interest rate on loans might now drive the worst projects rather the better ones out of the market; hence selection may be *favourable* rather than adverse from the banks' point of view. Also, if projects differ in terms of their expected returns, the analysis can lead to too much investment rather than too little as for the analysis under the assumption of common expected returns.

Furthermore, if projects differ in terms of their expected returns then clearly shares in them should not all sell for the same price; the use of equity is likely to lead to problems in this case. Indeed, since the owners or managers of firms have private information about their firms' expected returns, it may be those with the lowest expected returns who are most willing to sell their shares, thus leading to adverse selection problems in the equity market. Moreover, problems of an incentive type may occur with equity finance; because when a firm is equity-financed managers receive only a small fraction of any extra profit so their incentive to expend effort on making profits is attenuated.

To conclude, it seems that the market for investment finance is likely to suffer from a number of problems caused by asymmetries of information. These problems are likely to lead to complicated contracts governing the provision of finance and to the suppliers of funds expending much effort to try to learn about the projects for which entrepreneurs seek funds, or to monitor what entrepreneurs do with the funds they acquire. The next chapter looks in detail at the hidden action problem in the market for investment finance and shows once more that under some circumstances the use of equity finance avoids problems present under debt finance. Chapter 4 looks at the costly state verification problem in the market for investment finance and shows that in this case the standard debt contract will be used in the Nash equilibrium and will not be displaced by the use of equity finance.

2.6 Recommended reading

The seminal article on adverse selection is Akerlof (1970), which examines the market for used cars as an illustration of the problems that might be posed by asymmetric information.

An important article on asymmetric information in the credit market is Stiglitz and Weiss (1981); this deals with both the adverse selection and hidden action problems and so is useful for a background to both the present chapter and Chapter 3. Stiglitz and Weiss were concerned to use asymmetric information to explain the occurrence of credit rationing, although their analysis mainly made use of adverse selection in an example of the sort which our analysis above indicates should lead to equity rather than credit finance.

A survey paper dealing with asymmetric information and investment finance is Hillier and Ibrahimo (1993). Hillier and Ibrahimo (1992) generalises the analysis of this chapter by allowing projects to differ in terms of both their expected payoffs and the variance of their returns about their expected value. A good reference on the use of collateral requirements as signalling devices in the credit market is Bester (1985).

34   Investment Finance

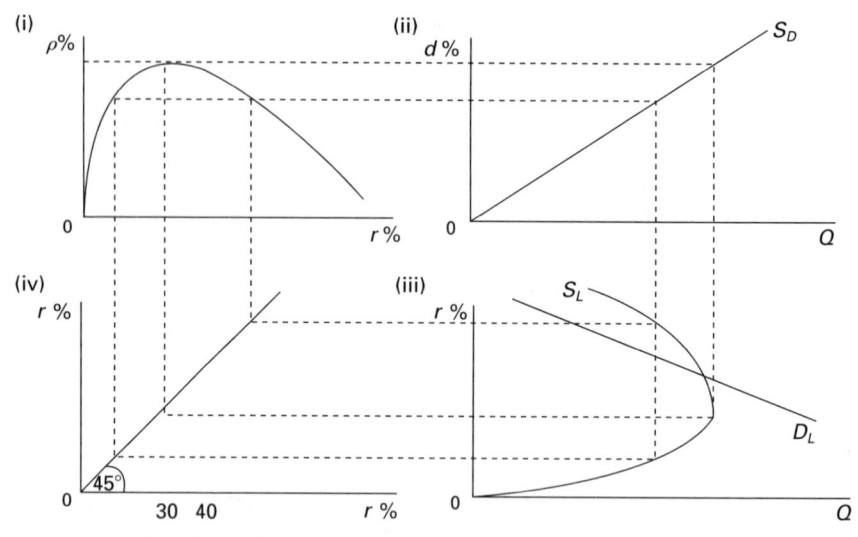

Figure 2.10 The credit market with a backward-bending supply of loans curve

2.7 Problems

Problem 2.1

Assume that the project types available, and their distribution among the population of entrepreneurs, is such that the relationship between ρ and r looks like that drawn in Figure 2.10(i). The relationship between ρ and r in Figure 2.10(i) is such that ρ rises as r is increased to r^*, and falls as r rises beyond r^*. Thus for interest rates above r^* the adverse selection effect dominates the beneficial effects of raising the interest rate.

Assume that all banks face the same relationship between ρ and r, and that they compete with one another for deposits.

(a) Explain the derivation of the loan supply function, which is the line showing the relationship between loan supply to the credit market and the interest rate charged on loans in Figure 2.10(iii).
(b) Figure 2.10 shows a downward-sloping loan demand curve, D_L. Why does the loan demand curve slope downwards?
(c) Will the credit market in Figure 2.10 exhibit credit rationing in equilibrium?
(d) Does a backward-bending supply curve always imply that credit will be rationed? Explain your answer carefully.
    (*Hint*: Reading Stiglitz and Weiss (1981) should help in answering this question.)

Problem 2.2
(a) Consider Figure 2.10 above. What would happen to the loan supply function, S_L, in Figure 2.10(iii) if the supply of deposits to the banking sector increased for any deposit rate, d?
(b) What would the movement in the loan supply function calculated in part (a) imply for the equilibrium interest rate and quantity of loans if both the old and new equilibria involve rationing?
(c) Your answers to parts (a) and (b) should show that the loan supply function can move without causing the loan demand curve to move. What factors would cause a move in the loan demand curve? Would they also be likely to move the loan supply curve?

CHAPTER 3

Investment Finance and the Hidden Action Problem

3.1 Overview

In this chapter we examine the implications of the hidden action problem in the market for investment finance. As in the previous chapter, we begin by considering that entrepreneurs can raise finance for their projects only by borrowing from banks. Section 3.2 shows that the problem of hidden action can lead to problems in the credit market. Section 3.3 shows how those problems can be solved by the use of equity finance for the example we present in Section 3.2. Section 3.4 offers a different example, where neither simple credit finance nor simple equity finance would solve the problem of hidden action. There is a possibility that the market for funds may collapse in this example. Possible solutions to the problem involving either punishment for fraud or costly monitoring of actions are suggested and discussed. Section 3.4 discusses the use of such amendments to equity finance in the example provided and Problem 3.1, with which the chapter ends, extends a similar analysis to the credit market. Section 3.5 looks at the way in which the hidden action problem may give rise to credit rationing in similar manner to the analysis of section 2.4 above.

3.2 Hidden action and the credit market

We assume, as in section 2.2 above, that entrepreneurs, banks and the suppliers of capital are risk neutral and that finance is provided under a standard debt contract.

Project returns

In the hidden action problem we assume that each entrepreneur can choose from several different investment projects. Banks are unable to observe the project in which an entrepreneur invests, so the act of investment is a hidden action.[1]

To focus attention, consider that there are n entrepreneurs, each of whom has a choice of two projects both costing K equal to 100 and both returning K if unsuccessful. An entrepreneur may invest in one of the two projects available to him but not in both of them. Type 1 projects return 150 if successful and have a 2/3 probability of success, while type 2 projects return 160 if successful and have a success probability of 1/2. The expected gross return to projects of type i (i equals 1 or 2) is given by the following equation:

$$E(R_i) = p_i R_i^s + (1 - p_i)K \tag{3.1}$$

where p_i and R_i^s represent the success probability and the payoff if successful for projects of type i. Using equation (3.1) yields values of 133.33 and 130 for the expected gross returns to type 1 and type 2 projects, respectively. Clearly, from a social point of view, it would be better for all investment to be in type 1 projects, since they yield a higher expected return than do type 2 projects for the same cost of investment.

The demand for funds

The entrepreneur's expected return from investing in a project may be calculated using equation (2.2) (see page 9). He will prefer to invest in type 1 projects as long as they yield a greater return than type 2 projects, which is the case as long as the following holds:

$$p_1[R_1^s - (1 + r)K] \geq p_2[R_2^s - (1 + r)K] \tag{3.2}$$

Constraint (3.2) simply says that the expected return for the entrepreneur is at least as large from investing in type 1 as in type 2 projects. We call constraint (3.2) the *incentive compatibility constraint*. Whatever contract the principal (the bank in our case) offers an agent (the entrepreneur in our case), the agent will then take the action that is best for himself. The principal will therefore be concerned to provide the agent with incentives to choose the action that is compatible with the principal's own objectives.[2]

After a little manipulation, constraint (3.2) yields the following:

$$(p_1 R_1^s - p_2 R_2^s)/(p_1 - p_2) \geq (1 + r)K \tag{3.3}$$

Using our example, constraint (3.3) holds for interest rates up to 20 per cent but does not hold for rates higher than this. The interpretation of this result is that for interest rates of up to 20 per cent, entrepreneurs would choose to invest in type 1 projects, but for rates higher than 20 per cent they will choose to invest in

Investment Finance

type 2 projects. The reader may verify this result by calculating the expected returns to the entrepreneur for both types of project, using equation (2.2), to see that for rates up to 20 per cent, type 1 projects yield entrepreneurs the higher expected return, and for rates greater than 20 per cent, type 2 projects yield the higher expected return. We thus see that the interest rate, r, acts not as a selection mechanism, as in the previous chapter, but as an *incentive mechanism* in the case of the hidden action problem, since it affects the actions taken by borrowers once they have obtained a loan.

The demand for loans schedule, D_L, is plotted in Figure 3.1 as the kinked function shown. The shape is easily explained with reference to the typical entrepreneur's participation constraint and the incentive compatibility constraint. Investment in type 1 projects would satisfy the participation constraint for any interest rate below 50 per cent, while type 2 projects satisfy the participation constraint for any interest rate up to 60 per cent. Hence, at any interest rate up to 60 per cent, each entrepreneur wishes to invest in a project, so the total demand for loans is nK, which equals $100n$ in our example.

The horizontal line, *ICC*, shown at an interest rate of 20 per cent is drawn to represent the incentive compatibility constraint: at rates above 20 per cent, investment will be in type 2 projects and at rates below 20 per cent investment will be in type 1 projects. At an interest rate of 20 per cent, entrepreneurs will be indifferent as to whether they invest in type 1 or type 2 projects. We assume that when indifferent, for selfish reasons, between two actions, the entrepreneur will take the action that is better from the point of view of the bank. Let us call this assumption *epsilon altruism*. This assumption implies a gain of a small amount, epsilon, for the agent when he takes an action that benefits the principal, but

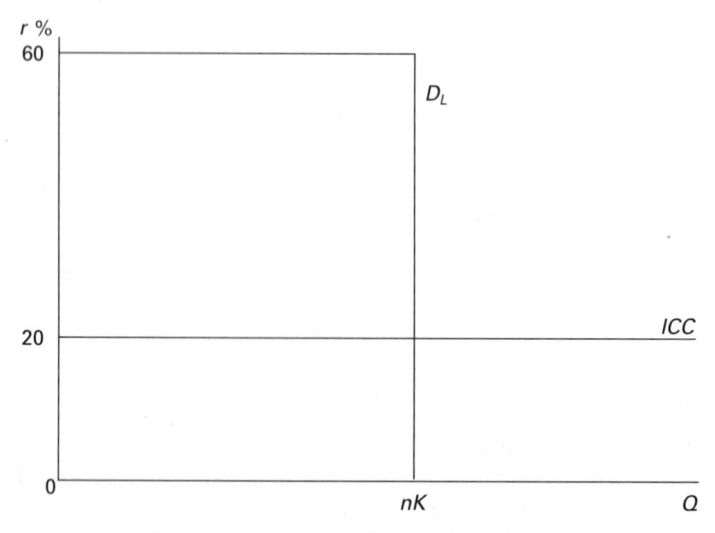

Figure 3.1   The demand for loans

that epsilon is so small that it cannot affect the agent's choice over actions whenever they offer different payoffs to the agent himself. We shall see below that, at an interest rate of 20 per cent, a bank would prefer entrepreneurs to invest in type 1 projects.[3]

The hidden action counterpart to adverse selection is known as *moral hazard with hidden action*.[4] The term *moral hazard* is applied because the actions taken by the borrowers are based on their own self-interest and not on the best interest of the lender. This term, like *adverse selection*, was first applied in the market for insurance, where it applies to any behaviour of the insuree that would work against the interests of the insurer; for instance, a motorist taking fewer anti-theft precautions once he has insured his car.[5] The extreme case of moral hazard is fraud; for example, a motorist arranging for his car to be 'stolen' so that he can claim compensation from an insurance company. Moral hazard occurs in the credit market if raising the interest rate induces borrowers, who have a choice of projects, to invest in a project that yields the bank a lower return than another project in which the borrower could have invested.

The supply of funds

As in the previous chapter, we consider the ultimate suppliers of funds to be private individuals who place their funds on deposit with banks. The analysis is facilitated if we assume, at least initially, that banks are able to attract as many deposits as they wish provided only that they offer a rate of return of d^* to depositors. As in Chapter 2 above, perfect competition in the banking industry then implies that banks will be willing to supply loans to entrepreneurs as long as the effective rate of interest on loans, ρ, is equal to d^*.

The $\rho - r$ relationship under moral hazard

The bank's expected gross return from lending to fund a project of type i is again given by:

$$E_b(\Pi_i) = p_i(1 + r)K + (1 - p_i)K = K(1 + rp_i) \tag{3.4}$$

which is just the probability of success times the repayment if successful plus the probability of failure times the repayment if unsuccessful. Hence the expected percentage return to the bank, ρ_i, from funding a given project is rp_i, which depends on the success probability of the project. Since we know that entrepreneurs choose to invest in type 1 projects for interest rates up to 20 per cent, and in type 2 projects for interest rates between 20 and 60 per cent, then we know that the expected value for ρ is given by:

$$\begin{aligned}\rho = rp_1 = (2/3)r & \quad \text{for } r \leq 20\% \\ \rho = rp_2 = (1/2)r & \quad \text{for } 20 < r \leq 60\%\end{aligned} \tag{3.5}$$

It is clear that at an interest rate of exactly 20 per cent the return to the bank is higher if the entrepreneur invests in his type 1 project, since it has a higher probability of success than his type 2 project. Given epsilon altruism, this explains why the entrepreneur chooses his type 1 project when r equals 20 per cent.

Equilibrium in the loan market

The relationship between ρ and r presented in equation (3.5) is shown in Figure 3.2, which plots ρ on the vertical axis and r on the horizontal. Each part of the figure shows two rays from the origin. The higher of these two rays, ρ_1, represents the effective return to the bank from lending to support type 1 projects, and the lower one represents the effective return from lending to fund type 2 projects; the higher ray therefore has a slope of 2/3 and stops at r equals 50 per cent, while the lower ray has a slope of 1/2 and stops at r equals 60 per cent.

The crucial difference from the analysis in the previous chapter is that now the market effective $\rho - r$ relationship is not an average of the two rays for low rates of interest. Instead, the market $\rho - r$ relationship for a bank unable to observe the project in which an entrepreneur places his funds is given by the discontinuous relationship $OPQR$. For rates of interest up to 20 per cent, the entrepreneurs would choose to invest in their type 1 projects so the returns to the banks are given along the upper ray for r values up to 20 per cent, and for rates greater than this they are given along the lower ray as entrepreneurs choose to carry out their type 2 projects.

The horizontal line at ρ equals d^* represents the required rate of return which banks must pay to depositors to attract funds, and is, therefore, the effective rate of return which must be made on loans in equilibrium.

Consider Figure 3.2(i). Imagine initially that banks can observe the project in which entrepreneurs place funds made available to them. The figure shows that in order to achieve an effective return of d^*, the bank could either charge a quoted loan rate of r_A on investment in type 1 projects, or a rate of r_B on investment in type 2 projects. For concreteness, assume d^* to be 12 per cent, then r_A equals 18 per cent and r_B equals 24 per cent. It is easy to see that in this case the entrepreneur would prefer to borrow at 18 per cent and invest in his type 1

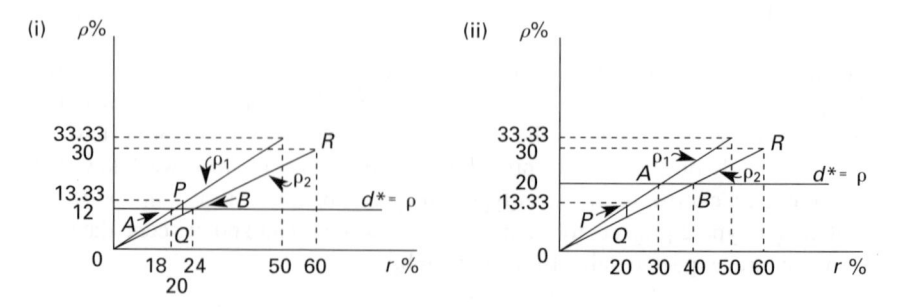

Figure 3.2 Credit market equilibrium

project than to borrow at 24 per cent and invest in his type 2 project. This is obvious from calculating explicitly the entrepreneur's expected returns in each case. Alternatively, note that the expected gross return for any project is divided between the entrepreneur and the bank. Hence, since type 1 projects have a higher expected gross return than type 2 projects, they must offer a higher expected return to the entrepreneur since the expected gross return to the bank is common for both types of project (that is, 112 in each case) for the two interest rates under discussion.

Introducing the asymmetry of information makes little difference in this case. Banks can still lend at a quoted loan rate of 18 per cent, safe in the knowledge that the incentive compatibility constraint is satisfied and entrepreneurs will choose to invest in their type 1 projects regardless of the asymmetry of information.

Now consider Figure 3.2(ii). Here the required rate of return for depositors is such that the horizontal line at d^* lies above the point P rather than below it as for Figure 3.2(i). For concreteness, let the value of d^* be 20 per cent. The asymmetry of information does matter in this case.

Consider first the full information case, where the banks can observe the project in which an entrepreneur invests. The analysis is then the same as above. The bank could obtain an effective rate of 20 per cent by charging a quoted rate of 30 per cent on investment in type 1 projects, or by charging 40 per cent on investment in type 2 projects. The bank's expected gross return on a loan of 100 in each case would be 120, leaving the remainder of the expected return to the entrepreneur. This remainder would be 13.33 if the entrepreneur borrowed to fund his type 1 project, or 10 if he borrowed to fund his type 2 project (that is, 133.33 minus 120, or 130 minus 120, respectively). The entrepreneur would therefore choose to borrow at the lower interest rate and to invest in his type 1 project.

Now, however, introduce the asymmetry of information. In this case, an entrepreneur borrowing at a quoted loan rate of 30 per cent would find it preferable to invest not in his type 1 project but in his type 2 project. This follows since the rate of 30 per cent is above the rate of 20 per cent, which we earlier determined was the rate at which an entrepreneur changed from choosing to invest in his type 1 project and would, instead, choose his type 2 project. The reader may calculate, using equation (2.2), that the entrepreneur's expected return from investing in his type 2 project is 15, which exceeds the expected return of 13.33 from investment in his type 1 project.

This increase in expected return to the entrepreneur is at the expense of the bank, which earns an effective rate of return of 15 per cent if it charges a quoted loan rate of 30 per cent on loans used to fund type 2 projects. In other words, in choosing his type 2 project rather than his type 1 project, the entrepreneur gains 1.66 but the bank loses 5 in terms of gross expected return, the net loss equalling 3.33, which is the reduction in expected gross return when comparing type 2 with type 1 projects.

42 Investment Finance

The bank, however, realises that the incentive compatibility constraint (3.5) is not satisfied at a quoted loan rate of 30 per cent. The bank is therefore not prepared to lend at that rate, since it knows it will make an effective rate of only 15 per cent and be unable to pay depositors their required rate of 20 per cent. Hence the bank will, instead, charge the higher rate of 40 per cent, knowing that this will induce the entrepreneurs to carry out their type 2 projects and produce an effective return of 20 per cent for the bank. Thus the equilibrium will be at point B rather than at point A in Figure 3.2(ii).

The problem with this equilibrium is, however, that it means that entrepreneurs are investing in their less productive project. The expected return to them is 10, which is less than the return of 13.33 that they could make if they borrowed at 30 per cent and invested in their type 1 projects. Thus entrepreneurs lose because of the asymmetry of information; if they could commit themselves to invest in their type 1 project if they were charged a quoted loan rate of 30 per cent they could gain 3.33 in expected returns, but their own tendency to invest in the project which yields them the higher return for a given quoted loan rate prevents them from being trusted by the banks. Thus, just like the selection problem in the previous chapter, it is possible that the hidden action problem can lead to a socially inefficient outcome in the credit market. The inefficiency in this case is that investment might be directed towards type 2 projects rather than towards type 1 projects, which offer a greater expected return for the same cost of investment.

3.3 The hidden action problem and equity finance

The problem with the use of credit finance in the face of the hidden action problem is that it may be necessary to set an interest rate so high that it gives entrepreneurs the incentive to invest in the wrong types of project. This problem can, at least for the example we have used so far, be overcome by the use of equity finance. The intuition for the use of equity finance is that since an entrepreneur under equity finance simply gains a percentage of a project's gross return he can be induced to carry out the project that has the greater expected gross return. This is exactly what the entrepreneur's shareholders would wish him to do, so there is compatibility between the objectives of the shareholders and entrepreneurs. In other words, equity finance leads to incentive compatibility at any share price in our example.

Consider how the equity market would be introduced instead of the credit market for the problematic case of Figure 3.2(ii). It could be introduced either by entrepreneurs, who could offer the banks just as good, or slightly better, an expected rate of return on funds provided by buying shares as on those provided by making loans while at the same time increasing the expected returns to themselves, or by banks who could offer to buy shares at such a price as to offer the entrepreneurs just as good a deal, or slightly better, as on the credit market while at the same time increasing the expected returns to themselves. Since, in either case, competition between banks forces the effective return on funds provided by them into equality with the rate they must pay the suppliers of funds, the equilibrium in the equity market will involve the extra returns going to the entrepreneurs rather than to the banks, so let us imagine that it is the entrepreneurs who actively seek to introduce equity rather credit finance.

Imagine a typical entrepreneur in Figure 3.2(ii). He is borrowing at a quoted loan rate of 40 per cent, choosing to carry out his type 2 project and making an expected return of 10. He could do better than this if he offered to sell shares in his type 1 project to the bank rather than borrow to fund his type 2 project. A bank requiring an effective rate of return of d^* per cent would be willing to pay V equal to $E(R_i)/100(1 + d^*)$ for a 1 per cent share in a project; for a type 1 project this yields $V = 1.33/1.2 = 1.11$. In order to fund his type 1 project the entrepreneur would therefore need to sell 90 shares (that is, 90×1.11 yields 100, which is the cost of the project). The entrepreneur would therefore keep a 10 per cent share in the return of his project. The expected return from share financing his type 1 project would thus be 13.33 (that is, 10 per cent of the gross expected return of 133.33), which is clearly better than the expected return of 10 from debt financing his type 2 project. Banks make an expected effective rate of 20 per cent from equity financing type 1 projects, just as from debt financing type 2 projects, therefore banks are willing to switch from credit to equity finance even though the gains from doing so go to the entrepreneurs.[6]

Replacing the credit market by the equity market allows a Pareto improvement to be brought about; the entrepreneurs gain and the banks (and the suppliers of funds) remain just as well off.[7] It is still necessary, however, to consider whether, having sold shares to the value of 100, the entrepreneurs would go ahead and carry out their type 1 rather than their type 2 projects. In order to see whether they would do so, it is necessary to examine the incentive compatibility constraint in more detail.

The incentive compatibility constraint may be written as follows:

$$E(\Pi_1) \geq E(\Pi_2) \tag{3.6}$$

which simply states that the expected returns to the entrepreneur from investing in his type 1 project are at least as large as those from investing in his type 2 project (given that he has acquired enough funds to fund either project).

The entrepreneur's expected returns from investing in the type 1 project are given by:

$$E(\Pi_1) = E(R_1)(S_1) = 133.33[(100 - 100/V)/100] \tag{3.7}$$

S_1 is the percentage equity stake retained by the entrepreneur, which is calculated by dividing 100 by V, to find the number of shares sold to raise the finance for the project, and subtracting this number from 100, to find the number of shares retained by the entrepreneur, and then dividing by 100 to transform the

44   Investment Finance

figure into a percentage. Multiplying the expected project gross return, $E(R_1)$, by S_1 yields the expected return to the entrepreneur from funding his type 1 project, which yields 13.33 in our example, where we know that V takes the value of 1.11, so S_1 equals 10 per cent.

The entrepreneur's expected returns from investing in the Type 2 project, given that he has sold enough shares to fund his type 1 project, are given by:

$$E(\Pi_2) = E(R_2)(S_1) \tag{3.8}$$

Notice that we continue to use S_1 in equation (3.8), since we assume the entrepreneur has sold shares to the value of 100 and retains the same equity stake whether he invests in his type 1 or type 2 project. For our example, $E(\Pi_2)$ is 13 (that is, 10 per cent of 130), which is less than the comparable figure of 13.33 for the type 1 project. Hence the incentive compatibility constraint is satisfied for the share price of 1.11. It would, in fact, be satisfied for any share price in this example, since both projects cost the same and the entrepreneur keeps the same equity stake regardless of which project he carries out; he would always therefore prefer to fund the project with the higher expected return, which is entirely compatible with the objectives of his shareholders. We shall see in the next section, however, that use of equity finance does not always guarantee incentive compatibility in this way.

The equity market equilibrium is a Nash equilibrium. No market participant can disturb it by offering a different arrangement. This follows since, from the equity market equilibrium, there is no offer any participant can make to any other participant that would simultaneously be attractive to both; that is, any move that benefited the entrepreneurs would harm the suppliers of funds and vice versa. Furthermore, the equity market equilibrium is also Pareto efficient; that is, it would be impossible to move away from the equilibrium without harming either the entrepreneurs or the suppliers of funds. Thus the use of equity finance has solved the problem of moral hazard with hidden action just as it solved the problem of adverse selection in the previous chapter.

Notice that in the case of Figure 3.2(i) there was no problem with the credit market. The reader ought to be able to work out that in this case there would be no gain in moving to an equity market, but that there would be a share price which could offer banks and entrepreneurs the same expected returns as the credit market. In this case, either the credit market or the equity market, or a mixed market with both equity and debt finance, would be possible Nash equilibria and all would be Pareto efficient.

3.4   Market collapse

Although the use of equity finance was able to overcome the problems posed by moral hazard with hidden action in the example above, it is possible to imagine examples where neither the use of equity nor credit finance can overcome the problem. Indeed, it is possible that the problem may be so severe as to cause the market for investment funds to collapse entirely. We present just such an example in this section and examine possible solutions to it. The use of equity finance is examined in the text. Problem 3.1 shows that the credit market would also collapse in this example unless bolstered by similar solutions as those discussed for the equity market.

As above, consider that there are n entrepreneurs, each of whom has a choice of two projects in which to invest and can invest in only one of them. Type 1 projects cost 100 and return 0 if unsuccessful, or 240 if successful, with equal probabilities of success or failure. Type 2 projects cost 50 and return 0 if unsuccessful or 100 if successful, with equal probabilities of success or failure. We assume that entrepreneurs have no funds of their own and must acquire them from the capital market if they are to fund their projects. The expected gross returns $E(R_i)$ to either type of project are given by the following equation:

$$E(R_i) = R_i^s/2 \tag{3.9}$$

where R_i^s is just the payoff if successful for projects of type i (i equals 1 or 2). Hence the expected gross returns are 120 and 50 for type 1 and type 2 projects, respectively. Thus, type 1 projects yield an expected rate of return of 20 per cent on the initial cost of 100, while the expected rate of return for type 2 projects is zero, given their initial cost of 50.

Imagine that the supply of funds is provided by an equity market. Assume that no funds would be supplied if they offered an expected return of zero to the suppliers, while funds would be supplied elastically for an expected return of 8 per cent to the suppliers of funds.

The full information case

Imagine initially that there is no problem of hidden action and that the suppliers of funds can observe the project an entrepreneur carries out. In this case it is clear that no funds would ever be provided for type 2 projects, since the suppliers of funds are not willing to supply funds for an expected return of zero. On the other hand, enough funds would be willingly supplied to fund each entrepreneur's type 1 project at an expected rate of return of 8 per cent to the suppliers of funds. At this expected rate of return, share prices would be given as follows:

$$V = E(R)/(1 + d) = (1.2)/(1.08) = 1.11 \tag{3.10}$$

The number of shares sold per entrepreneur at this price would be 90 (that is 100/1.11), leaving entrepreneurs a 10 per cent stake in their projects and an expected gross return of 12 from carrying them out (the remaining expected gross return of 108 going to the suppliers of funds). Thus, under full information, each entrepreneur is able to invest in his type 1 project while no funds would ever be made available for type 2 projects which offer an expected return of zero.

46 Investment Finance

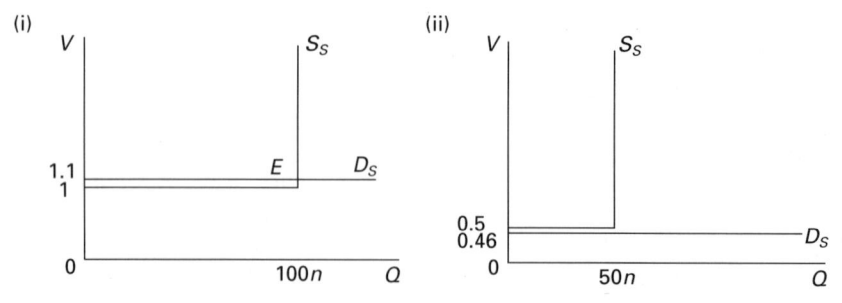

Figure 3.3 **Equity markets under full information**

The equity markets in the full information case are illustrated in Figure 3.3, where Figure 3.3(i) shows the market for shares in type 1 projects, and Figure 3.3(ii) shows the market for shares in type 2 projects. The kinked shapes of the supply schedules are explained by the participation constraints of the entrepreneurs, who would have nothing to gain by offering to sell shares at prices below 1 for type 1 projects, or below 1/2 for type 2 projects (since they only cost 50 to carry out). The demand curves are consistent with a perfectly elastic supply of funds schedule at an expected rate of return of 8 per cent. Since the supply and demand curves do not intersect in Figure 3.3(ii), it is clear that in this case no funds would be supplied for type 2 projects.

Moral hazard and market collapse

Now consider the case where the suppliers of funds are unable to observe which project entrepreneurs carry out. In this case it is necessary to consider the entrepreneur's incentive compatibility constraint to see if he would actually choose to invest in his type 1 project once he had sold enough shares to fund it.

We assume, initially, that once an entrepreneur has sold shares he can invest in whichever project he chooses and pay to the other shareholders the proportion of the eventual gross return that their shareholding specifies. Thus if, as in the full information case above, the suppliers of funds provide funds by buying shares at a price of 1.11 and by so doing provide funds of 100 to an entrepreneur in exchange for a 90 per cent shareholding in future gross returns, the entrepreneur can then decide to invest in either his type 1 or his type 2 project and keep a 10 per cent equity stake regardless of which project is carried out. If he invests in his type 2 project he can keep, or embezzle, the saving in investment cost of 50. The entrepreneur's expected return from investing in his type 2 project is therefore equal to 55; that is, the 50 saving in investment cost plus his 10 per cent stake in the project's expected gross return of 50. Since this expected return is much larger than the expected return of 12 from carrying out his type 1 project, the entrepreneur has an incentive to invest in his type 2 project rather than his type 1 project.[8] The full information outcome, then, is clearly no longer achieved under asymmetric information. Indeed, in this example the market collapses entirely, as may be shown by considering the entrepreneur's incentive compatibility constraint in more detail.

The incentive compatibility constraint may be written as follows:

$$E(\Pi_1) \geq E(\Pi_2) \qquad (3.11)$$

which simply repeats constraint (3.6) and states that the expected returns to the entrepreneur from investing in his type 1 project are at least as large as those from investing in his type 2 project, given that he has acquired enough funds to fund the more costly type 1 project.

The entrepreneur's expected returns from investing in the type 1 project are given by:

$$E(\Pi_1) = E(R_1)(S_1) = 120[(100 - 100/V)/100] \qquad (3.12)$$

S_1 is the percentage equity stake retained by the entrepreneur, which is calculated by dividing 100 by V, to find the number of shares sold to raise the finance for the project, and subtracting this number from 100, to find the number of shares retained by the entrepreneur, and then dividing by 100 to transform into a percentage. Multiplying the expected project gross return, $E(R_1)$, by S_1 yields the expected return to the entrepreneur from funding his type 1 project.

The entrepreneur's expected returns from investing in the type 2 project given that he has sold enough shares to fund the costlier type 1 project are given by:

$$E(\Pi_2) = E(R_2)(S_1) + 50 = 50[(100 - 100/V)/100] + 50 \qquad (3.13)$$

Notice that we continue to use S_1 in equation (3.13), since we assume the entrepreneur has sold shares to the value of 100 and retains the same equity stake whether he invests in his type 1 or type 2 project. However, since type 2 projects cost only 50, the entrepreneur is able to keep 50 for himself if he invests in his type 2 project.

Substituting from equations (3.12) and (3.13) into inequality condition (3.11) yields the incentive compatibility constraint as follows:

$$120[(100 - 100/V)/100] \geq 50[(100 - 100/V)/100] + 50 \qquad (3.14)$$

Manipulation of constraint (3.14) shows that it holds only for values of V greater than or equal to 3.5. Since this value for V is above the highest feasible value of 1.2 (at which price the suppliers of funds would expect to make a zero return on equities) it follows that the incentive compatibility constraint is never satisfied in this model. Entrepreneurs would therefore always have the incentive to invest in type 2 rather than in type 1 projects if they received funds. Knowing this, the suppliers of funds would refuse to purchase equities, since they do not wish to fund type 2 projects.

48   *Investment Finance*

The equity market would collapse and no investment would take place, even though opportunities existed to invest in type 1 projects to the advantage of both entrepreneurs and suppliers of funds. These opportunities are not grasped because the problem of hidden action would induce entrepreneurs to invest in type 2 rather than type 1 projects if they could acquire funds.

Figure 3.4 illustrates the collapse of the equity market. The supply and demand curves are drawn as for Figure 3.3(i) above, since we are looking at the market for shares in type 1 projects. The difference between Figure 3.3 and Figure 3.4 is that the latter shows the incentive compatibility constraint, *ICC*, as the horizontal line at a share price of 3.5. Since the demand curve for shares lies below the *ICC* line, we know that investment would be in type 2 and not type 1 projects if shares were bought at those prices; incentive compatibility is not satisfied at any price for shares that the suppliers of funds would be willing to pay, and the market collapses.

A legal solution

The collapse of the share market is undesirable, since it prevents entrepreneurs and the suppliers of funds from making feasible gains in expected returns. Since we have assumed that the returns to investment are observable it is possible to devise a scheme to ensure that entrepreneurs have an incentive to carry out their type 1 projects. This is possible since, if we observe a return of 100, we know that the entrepreneur invested in his type 2 project rather than in his type 1 project.[9]

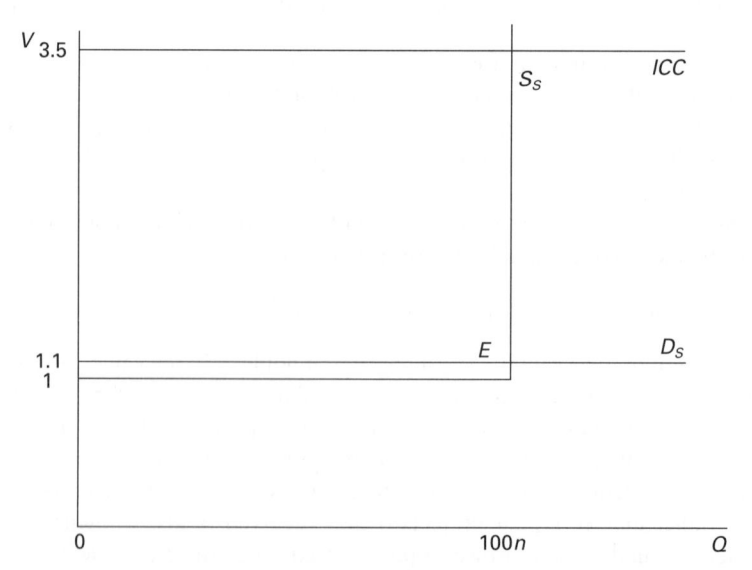

Figure 3.4   Collapse of the share market

If entrepreneurs are punished whenever a return of 100 is observed, showing that they invested in their type 2 projects, then if the punishment is big enough it is possible to ensure that the incentive compatibility constraint is always met. This is done by making $E(\Pi_2)$ sufficiently small by subtracting from equation (3.13) the expected value of the punishment. If an entrepreneur invests in his type 2 project there is a 50 per cent chance of it producing a payoff of 100, which reveals that he invested in it rather than in his type 1 project, and so a 50 per cent chance of him being punished. If this punishment is big enough he will not invest in his type 2 project for fear of being punished. Implementing this policy, or legal solution of punishing embezzlers, ensures that the incentive compatibility constraint is satisfied and the share market can operate to provide funds for type 1 projects to the advantage of all concerned.[10]

Pour encourager les autres?

Although the punishment policy solves the problem of market collapse, and helps to explain why laws against fraud are common (similar arguments explain why a reputation for honesty is often argued to be a good business asset) it is not as easy to implement as might first appear. The difficulty is that the punishment necessary to make the policy work may be quite severe. In this case, although the policy works if entrepreneurs believe the punishment will be inflicted if they embezzle, some entrepreneurs may believe that no judge would ever impose such a harsh penalty on them and so may decide to invest in type 2 projects and embezzle funds despite the announced punishment. If this happens, judges, and society more generally, face a severe dilemma – should the penalty be imposed, which may seem to be a harsh and uncivilised thing to do, or should a lesser penalty be imposed and the potential gains the policy sought to produce placed at risk?

Monitoring

If the punishment policy is not to be used it might be possible, although at some expense, for the suppliers of funds to monitor the actions of the entrepreneurs to ensure that they invest in their type 1 projects. Since such monitoring is costly, it reduces the gains from funding investment, but as long as it is not too costly it is better than a market collapse. The need for monitoring of this type may help to explain why institutional suppliers of funds might wish to have representatives on the boards of companies in which they invest.

Another problem with the punishment policy is that it only works so well in our example because the payoff of 100 reveals quite clearly that the type 2 project had been carried out. The policy works in that case because the two project types have quite different payoff distributions.[11] If the payoff distributions were more complicated, say both projects produce payoffs from continuous distributions which largely overlap one another, then it may be difficult, by observing the payoff, to deduce which project had been carried out. It therefore becomes difficult to devise a suitable punishment scheme without risking punishing the innocent. In such a case there is even more reason to resort to monitoring actions rather than to seek a punishment solution.

3.5 Credit rationing

Consider once more the example of section 3.2 above, where type 1 projects have a 2/3 chance of success with an associated payoff of 150, and type 2 projects have a 1/2 chance of success with an associated payoff of 160. Both types of project cost 100 and yield 100 if unsuccessful.

The supply of deposits

The hidden action problem may be used to explain credit rationing in much the same way as the selection problem was used in the previous chapter. As with the selection problem, it is necessary simply to introduce the assumption that the supply of deposits placed with banks increases as the rate of interest, d, paid to depositors increases. The supply of deposits schedule, S_D, is again given as shown in Figure 2.5 on page 23.

The supply of loans

The supply of loans, S_L, may now be derived using Figure 3.5. Figure 3.5(i) of the figure shows the $\rho - r$ relationship presented in equation (3.5) above.

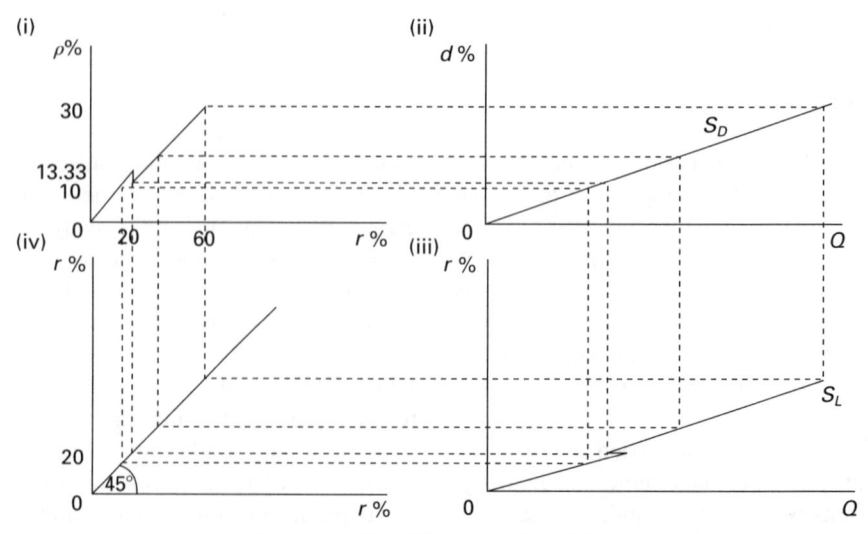

Figure 3.5   The supply of loans

Figure 3.5 may be interpreted in similar manner to Figure 2.6 on page 23. Consider some interest rate, r, say 40 per cent, in Figure 3.5(i) and read up to the $\rho - r$ relationship to find the corresponding value of ρ. Reading across from the ρ value in Figure 3.5(i) to the equal value for d in Figure 3.5(ii) (which we can do, since in equilibrium ρ will be equal to d) we can then read down to find the volume of deposits, Q, that could be attracted by the banking industry for the interest rate, r, with which we started. Since these deposits can be supplied as loans by the banks we can then plot the volume of loans, Q, against the interest rate, r, in Figure 3.5(iii).

Notice that the supply of loans so derived is discontinuous at an interest rate of 20 per cent. This is because of the discontinuity in the $\rho - r$ relationship at that interest rate, which also shows up in the supply of loans function. The intuition is straightforward. As the interest rate passes 20 per cent, the incentive for entrepreneurs to switch from investing in type 1 projects to type 2 projects causes the average rate of return to the banks to fall and hence they can attract fewer deposits. As the interest rate rises further above 20 per cent, the average rate of return to the banks rises once more and they begin to attract more deposits again. The supply curve, like the $\rho - r$ relationship, stops at $r = 60$ per cent, since at interest rates greater than 60 per cent, entrepreneurs no longer seek funds.

Equilibrium in the credit market

It is now possible to consider the equilibrium in the loan market using Figure 3.6, which plots the supply and demand for loans curves on the same diagram. The figure shows three possible outcomes.

Figure 3.6(i) shows the case where the supply and demand curves intersect twice (ignoring the intersection on the discontinuous part of the supply schedule), at points A and B in the figure.

Both points A and B represent points where supply equals demand, but B will unambiguously be the market equilibrium. This can easily be seen, since at B the supply of funds is the same as at A, which means that the average rate of return to the banks, ρ, and the deposit rate, d, must be equal at A and B. However, the return to entrepreneurs must be greater at B, since at B they are investing in type 1 projects, which yield a greater gross expected return than the type 2 projects they would choose to invest in if the economy operated at point A. Since the expected return to the bank is the same at A or B, then the expected return to the entrepreneur, which is just the remaining part of total returns, must be greater at B. Therefore, starting from point A, a bank could ask for an interest rate significantly below A but above B, say a rate of r_C (shown at point C in the diagram) and attract entrepreneurs away from A while increasing bank profits. The rate r_C must be chosen to be above r_B but below the critical rate of 20 per cent (at which the entrepreneur's incentive compatibility constraint holds with equality) if it is to increase the bank's expected profits; otherwise the entrepreneurs will continue to invest in their type 2 projects at the new lower interest,

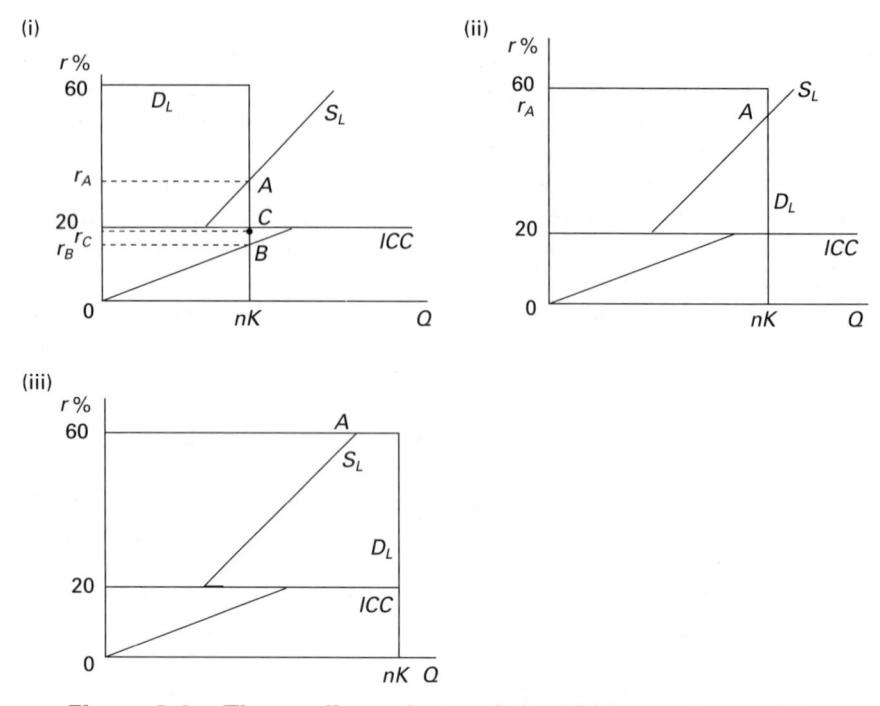

Figure 3.6 The credit market and the hidden action problem

rate and the bank's expected profits would decline. However, as long as r_C is above r_B and below 20 per cent, both banks and borrowers gain compared to the starting point of A: the entrepreneurs gain from paying a lower interest rate and the banks gain because entrepreneurs now choose their type 1 projects and so default on their debts with lower probability, which more than offsets the effect on bank profits of the reduction in the quoted interest rate. This could fall as far as r_B before these two offsetting effects exactly matched one another.

The interest rate of r_C does not produce a Nash equilibrium, since the expected effective return to the banks at this rate would exceed the expected return necessary to attract the amount of funds $100n$ as deposits, since point C is above the supply of loans function. Thus banks make supernormal profits at C (that is, p exceeds d). Competition in the banking sector, therefore, would drive down the interest rate charged on loans until the supernormal profits are eliminated at B with an interest rate r_B.

In Figure 3.6(i) the asymmetry of information does not create a problem. At B, entrepreneurs choose to invest in type 1 projects, which yield a higher expected return than type 2 projects and are the types of project that should, therefore, be funded. The reason for the satisfactory outcome is simply that the supply of funds, which underlies the derivation of the supply of loans schedule, is such that all the demand for loans can be met at an interest below 20 per cent. Hence entrepreneurs choose to invest in type 1 projects, which is the socially desirable outcome.

Unfortunately, as the next two cases show, this satisfactory state of affairs does not persist if the supply of funds is reduced relative to the demand for them.

Figure 3.6(ii) shows a more problematic case. Here the supply of funds is relatively less abundant and there is a unique intersection of the supply of loans function with the demand for loans function at A for an interest rate, r_A, greater than 20 per cent. The credit market equilibrium therefore occurs at A. All entrepreneurs receive funds to finance their projects but now, since the interest rate is above 20 per cent, they all choose to invest in type 2 projects. Since the expected return for type 1 projects is greater than that for type 2 projects this outcome is clearly inefficient. If entrepreneurs could only be induced in some way to invest in type 1 projects rather than type 2 projects, it would be possible for the gains available from doing so to be shared among the entrepreneurs and suppliers of funds in such a way that either one or both groups could gain. The credit market is, however, unable to secure the available gains and produces an inefficient outcome because of the asymmetry of information. We shall see below that, as in section 3.3, this problem may be overcome, for our specific example, by the use of equity rather than credit finance.

Figure 3.6(iii) shows the credit rationing case. Here the supply of funds is even more scarce relative to demand. The unique intersection of the supply and demand schedules occurs at the maximum interest rate possible for this market and at a point A on the horizontal part of the demand schedule. Thus the interest rate is 60 per cent and the demand for loans exceeds the supply, so there is rationing. However, with the interest rate at 60 per cent, entrepreneurs are indifferent whether they invest in type 2 projects or not at all, so it does not seem that the rationing of funds should be considered a problem. However, as for the analysis of Figure 3.2(ii), there *is* a problem. One aspect of the problem is that the investment that does take place is in type 2 rather than type 1 projects, which yield a higher expected return. Furthermore, if entrepreneurs could only commit themselves to investing in type 1 projects, the extra expected return could not only be shared among the entrepreneurs and the suppliers of funds in such a way as to make them all better off, it would also ease the rationing of funds, since a higher return to the suppliers of funds would attract an increased supply. Hence, more funds could be made available if the return to their suppliers was increased, and yet projects capable of offering such returns are not being funded: the level of investment is, therefore, inefficiently low. Unfortunately, entrepreneurs have an incentive to invest in type 2 projects for any interest rate over 20 per cent. Any commitment to invest in type 1 projects would not be believed by the lenders of funds, since they cannot observe the type of investment made and assume that each entrepreneur will always invest in whatever project yields the higher expected return.

Equity finance

As for the simpler version of our example in sections 3.2 and 3.3 above, where the supply of funds is perfectly elastic, the problems posed by hidden actions

54   Investment Finance

may be solved by using equity finance for projects of the type we are considering here, even though we now have a less than perfectly elastic supply of funds schedule.

The problem with the use of credit to finance investment projects in the face of the hidden action problem is that as the interest rate on loans rises it induces entrepreneurs to invest in the 'wrong' type of project. Hence the problem in Figure 3.6(ii) and (iii), that entrepreneurs invest in type 2 rather than type 1 projects. The interest rate, or more generally the credit contract, may provide entrepreneurs with the wrong incentives in these cases. In the case of Figure 3.6(i) there is no problem, because the supply of funds is such that all type 1 projects can be financed at an interest rate below 20 per cent. In this case the entrepreneurs' actions, though still hidden, will not cause a problem, since they choose to invest in type 1 projects in any case.

The use of equity finance may be shown to solve the hidden action problem in our example. The basic reason for this is that, in this example, equity finance always provides entrepreneurs with the incentive to invest in type 1 projects. The need for a high return to attract an increased supply of funds is met in the equity market by reducing the price of shares which, unlike a rise in the interest rate, does not induce entrepreneurs to switch to type 2 rather than type 1 projects. Furthermore, it is easy to see that, in cases like those illustrated in Figure 3.6(ii) and (iii), the credit market does not produce a Nash equilibrium. Competitive forces will bring about an equity market and produce a Nash equilibrium.

Let us consider Figure 3.6(ii); similar arguments also follow for Figure 3.6(iii). Imagine, for concreteness, that r_A is 30 per cent. The incentive compatibility constraint implies that entrepreneurs invest in type 2 projects and it then follows, from equation (3.5), that banks receive an expected return of 15 per cent. Could a bank in this situation do anything to improve its profits, given the behaviour of other banks, depositors and entrepreneurs? The answer to this question is yes. The bank could offer to buy an equity stake in the type 1 projects of entrepreneurs rather than lending to them. Using equation (2.9) (page 17) we can see that, in order to make an average return of 15 per cent, the bank would be able to pay a price per share approximately equal to 1.159; this is calculated by setting the value of d equal to 15 per cent and the value of $E(R)$ equal to 133.33 in the formula, V equals $E(R)/100(1 + d)$.

Entrepreneurs selling shares at a price of 1.159 and then investing in their type 1 projects would make an expected gross return from so doing of 18.33. This may be calculated by noting that they need to sell 86.25 shares to fund the project (that is, 100 divided by 1.159) and so retain a personal equity stake of 13.75 per cent in their projects, which yield them an expected gross return of 18.33 (that is, 13.75 per cent of 133.33). This return compares favourably with the expected gross return of 15 available from borrowing at 30 per cent to invest in type 2 projects (which may be calculated using equation (2.2) on page 9).[12]

Thus the entrepreneur would be attracted to a bank offering to buy shares at a price of 1.159 rather than to lend to him.

Since we assume that the bank cannot observe the investment project undertaken by the entrepreneur and he could invest in his type 2 project if he so chose, we must also consider whether the entrepreneur, having sold shares, would choose to invest in the type 1 project rather than the type 2. We know already, from the arguments in section 3.3, that the incentive compatibility constraint is satisfied at any share price for this example. Hence, we know that entrepreneurs would, indeed, carry out their type 1 projects if they received equity finance. The logic for this is quite straightforward and just repeats the arguments of section 3.3. Entrepreneurs, having sold shares, will choose to invest in type 1 projects only if doing so yields them a higher expected return than investing in type 2 projects. Since both types of project are equally costly, with K equal to 100, their equity stake is equal to 13.75 per cent regardless of which project they carry out. Therefore entrepreneurs always prefer to invest in the project with the higher expected return; that is, in type 1 projects. The suppliers of funds similarly would prefer investment to be in type 1 projects under equity finance. Thus switching from a credit market to an equity market solves the problem of incentive compatibility; in this example, both suppliers of funds and entrepreneurs always prefer type 1 to type 2 projects under equity finance.

Thus, starting from a credit market regime with an interest rate of 30 per cent, a bank could offer to buy shares at a price up to 1.159 and attract entrepreneurs away from banks offering credit. The first bank to come up with this idea could make supernormal profits by paying a price less than 1.159 but high enough to attract entrepreneurs. The supernormal profits would, however, be noticed by other banks, who would compete it away by copying the innovator and buying shares. The credit market regime would be replaced by an equity market equilibrium and competition between the buyers of shares would push the share price up to 1.159 at the new equilibrium. The price in the new equilibrium must be 1.159, since at this price the return to the suppliers of funds is 15 per cent, which we assumed in the credit market case was just sufficient to attract enough funds to the market to fund each entrepreneur's project. In other words, the competition constraint still holds as for the analysis of the credit market.

In the equity market equilibrium the suppliers of funds make an expected return of 15 per cent, therefore they neither lose nor gain as a result of the move from a credit to an equity market regime, even though it is the possibility of supernormal profits for the first bank to innovate by offering equity finance which we assumed led to moving the economy from the credit market regime in the first place (alternatively, we could have allowed the initiative to be taken by entrepreneurs, as we did in section 3.3). The move from the credit to the equity market is, however, clearly a Pareto improvement, since entrepreneurs gain an increase in their expected returns from 15 to 18.33 as a result of the move to equity finance.

The reader ought to be able to work out that, just as in Section 3.3 above, the equity market equilibrium is a Nash equilibrium and is also Pareto efficient.

56 Investment Finance

Similar arguments to those presented above may be applied to the credit market in Figure 3.6(iii). The major complication concerns the possibility of rationing of funds in the equity market in this case. This possibility arises because in Figure 3.6(iii) the return to the suppliers of funds is 30 per cent and at that rate the supply of funds is insufficient to fund all projects. Therefore, even under equity finance there could be rationing if the supply of funds at the maximum possible rate of return to the providers of funds under the equity market (which is 33.33 per cent at a share price of 1 for shares in type 1 projects) is less than the demand for funds (or the supply of shares) at that rate of return. Any rationing that took place would be determined in an arbitrary manner and, as in the previous chapter, we assume that entrepreneurs either sell enough shares to fund their projects, or sell none at all. Furthermore, and again as in the discussion of equity rationing in the previous chapter, any equity rationing in equilibrium would be efficient, since the rationing in this case would not be caused by asymmetric information but rather by an insufficient supply of funds at the maximum return possible from projects: not all projects should be funded, since they do not offer a sufficient return to cover the opportunity cost of funds.

The equity market may be illustrated in exactly the same way as for the equity market case in the previous chapter, using Figure 2.9 on page 29. The only points to note are that the demand and supply schedules are both drawn with respect to type 1 projects only in this case, and that the maximum value of V for a value of d equal to zero is now 1.33 rather than 1.2 as in the previous chapter.

3.6 Problem

Problem 3.1
Consider the moral hazard problem of section 3.4.

(a) If the funds for type 1 projects were provided via a credit market, what rate of interest would need to be charged to entrepreneurs under a standard debt contract if the suppliers of funds are to receive an expected rate of return of 8 per cent?

(b) Assume that the act of investment can be observed, but that the lenders cannot tell in which project the entrepreneur invests, and that if he invests in the type 2 project he is able to keep the saving on investment costs. Would an entrepreneur who borrowed funds of 100 under the debt contract calculated in part (a) above choose to invest in his type 1 or his type 2 project? Would a punishment policy induce the entrepreneur to invest in his type 1 project? Explain your answer by showing how the punishment policy affects the incentive compatibility constraint.

CHAPTER 4

Investment Finance and the Costly State Verification Problem

4.1 Overview

In this chapter we examine the implications of the costly state verification problem in the market for investment finance. Section 4.2 shows that, in the face of this problem, finance is optimally intermediated via banks and that the standard debt contract may be derived as the optimal contract. As in the previous chapters, it is possible that the credit market may be characterised by rationing in equilibrium; the policy implications of such an outcome are discussed. Section 4.3 shows how the costly state verification problem may be used as the basis of a macroeconomic model that offers an explanation for the persistence of business cycle shocks. Section 4.3 is something of a diversion from the more microeconomic thread of the rest of the book and may be skipped if the reader wishes, although it is useful in showing how asymmetric information may have important macroeconomic implications. Section 4.4 suggests some reading and section 4.5 presents some problems.

4.2 Hidden information and the credit market

Consider that there are n entrepreneurs, each of whom is endowed with one project. Assume that the return to each project is a random variable, R, and all entrepreneurs who invest receive a return drawn from the same probability distribution, all projects cost K and each entrepreneur needs to acquire funds to the value of K before he can invest in his project. Let K equal 100.

Assume that R will be determined by a drawing from a uniform distribution. For such a distribution, R can lie anywhere between a lower amount, a, which is the worst return the project can yield, and an upper amount, b, which is the best return the project can yield. Before the return is known, the entrepreneur

f(R)

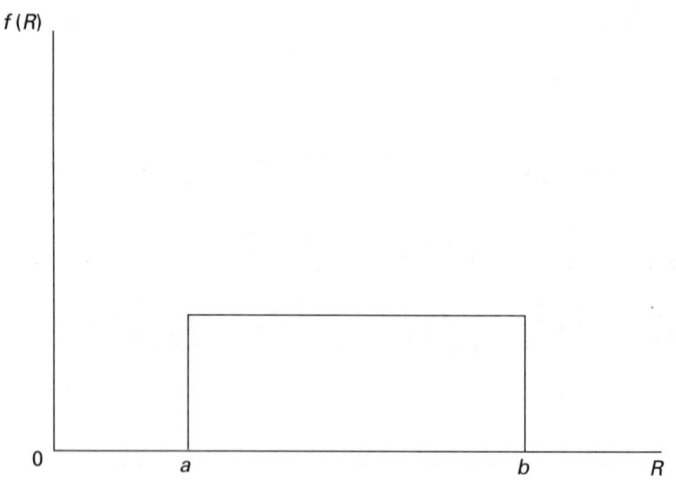

Figure 4.1 A uniform density function

attaches an equal likelihood to obtaining any value within the limits a and b; the density function, therefore, looks like that shown in Figure 4.1. The average or expected return, $E(R)$, on such projects is given by the amount $(a + b)/2$, and the density function is given by $f(R) = 1/(b - a)$. Let a equal 100 and b equal 200, so that the expected return is 150.

Assume that the bank, or supplier of funds, knows the probability distribution from which project returns will be drawn, but cannot observe the actual return on any specific project without incurring a monitoring cost of c equal to 10. Thus before the project yield is drawn there is no information asymmetry and all projects and entrepreneurs are equivalent from the point of view of the suppliers of funds. The information asymmetry only arises once projects have yielded their returns.

Let us now see how the standard debt contract, with monitoring of projects unable to repay the loan, arises as the optimum response to this situation. We shall present the argument by beginning with a share contract and showing how competition would cause banks to change the contract terms, in search of supernormal profits, until we arrive at a Nash equilibrium in which banks offer the standard debt contract. Throughout we shall assume that the suppliers of funds will provide enough funds to support all available projects, provided they receive an expected return of 25 per cent. Furthermore, we assume that funds are channelled from the suppliers of funds to entrepreneurs via financial intermediaries, which we call banks.

The share contract with monitoring

One feasible contract would be for a bank to fund a project in return for a 90 per cent share of whatever return it yields. The moral hazard with hidden information problems means that an entrepreneur faced with this contract, and knowing that the bank cannot observe the project return, has the incentive to announce a low return even if the true return is high. To ensure that it receives its specified share of the return to the project a bank must therefore commit itself to monitor the project, to observe the return.[1] Let us call the resulting contract 'the 90 per cent share contract with monitoring'. The expected return to the bank net of monitoring costs will then be 90 per cent of the gross expected project return minus the monitoring cost; that is, (0.9×150 − 10), or 125. Hence, a bank funding a large number of projects under such a contract would make an average return net of monitoring costs of 25 per cent. The banking sector could offer such contracts to entrepreneurs and pay the suppliers of funds the deposit rate of 25 per cent necessary to attract enough funds to support all projects. The 90 per cent share contract with monitoring is therefore feasible and could be used to channel funds from the suppliers of funds to entrepreneurs. However, it is easy to show that such a contract will not produce a Nash equilibrium.

The critical value contract

Under the 90 per cent share contract with monitoring, banks break even; they earn an average rate of return of 25 per cent and pay the same rate to depositors. It is easy, however, for a bank to change its behaviour and gain supernormal profits if all other banks are offering the 90 per cent share contract and monitoring all borrowers. All that a bank needs to do is to offer entrepreneurs a 90 per cent share contract but commit to monitoring only when an entrepreneur's announced payoff is less than a critical value of 189. Let us call such a contract a 'critical value contract'. Clearly, entrepreneurs would prefer such a contract to the 90 per cent share contract with monitoring, since if their projects yield more than 189 they do not need to reveal this to the bank but can, instead, claim a return of 189 and keep all the extra returns (rather than just 90 per cent of them).

The critical value is arrived at by finding that project return which yields a payoff to the bank when it is not monitored which is equal to the maximum payoff under the share contract with monitoring of all projects. This may be found by using the following formula:

$$CV = (sb - c)/s = b - c/s \tag{4.1}$$

where CV is the critical value, b is the maximum project payoff, c is the monitoring cost, and s is the share of the returns taken by the bank. Using the values for our example, the maximum payoff to the bank from a single project with monitoring is $(0.9 \times 200$ minus $10)$, which equals 170 − that is, 90 per cent of 189. Choosing not to monitor when 189 is announced is, therefore, as good as monitoring and finding the actual return to be the maximum of 200. For any project yielding less than the maximum return but more than 189, it is clearly not worthwhile to monitor, since the monitoring cost exceeds the extra share the bank can claim from finding the true value of the return. The bank can, therefore, make profitable savings on monitoring costs by not monitoring whenever an entrepreneur announces a return of 189 or more.

60   *Investment Finance*

For projects with an announced return of less than 189, it is possible that it would be worthwhile to monitor, since it might be the case that the monitoring cost could be exceeded by the extra return to the bank if the entrepreneur was found to be lying. Under the new contract the bank receives the same payoff on projects which produce a return of less than 189 as it would under the contract with monitoring of all projects, since it monitors these projects under either contract. The gains from savings on monitoring costs on the more successful projects under the new contract therefore represent an increase in profits for the bank compared to the share contract with monitoring of all projects. The share contract with monitoring will not, therefore, produce a Nash equilibrium, since banks would replace it by a critical value contract.

The Revelation Principal and maximum payment

Under the critical value contract an entrepreneur knows that the bank will not monitor if he announces a return of 189, but that if she obtains and announces a return greater than 189 the bank will take 90 per cent of that return from him. Therefore he will never announce a return greater than 189, since to do so involves volunteering to make a greater payment than necessary to the bank. The bank, however, realises that entrepreneurs with projects yielding a return greater than 189 will respond to the critical value contract by lying about their project return in this way. The bank can therefore change the contract terms so that it asks for a 90 per cent share of the project return up to a maximum payment to the bank of 170, that is 90 per cent of 189, without affecting the actual payments to the bank. The effect of the change is simply to remove the need for entrepreneurs with very successful projects to understate their returns.

The change from the critical value contract to the critical value contract with a maximum payment, which we shall call the maximum payment contract, illustrates a useful idea known as the *Revelation Principle*. The Revelation Principle states that for every contract that leads to lying (that is, in our example, to understating the value of project returns) there is another contract with the same payoff to the principal for any outcome (that is, in our example, the same payoff to the bank for any project return) but no inducement for the agent (the entrepreneur) to lie.

For the Revelation Principle to hold, we assume that the agent will tell the truth if he receives the same payoffs from telling the truth or from lying. This latter assumption is known as *epsilon truthfulness*. Epsilon truthfulness implies that the agent gains a small amount, epsilon, from telling the truth rather than lying, but that he will choose to lie whenever the payoff from lying exceeds that from telling the truth.

The value of the Revelation Principle is that it allows us to examine only contracts which induce truth-telling. We can thus impose *truth-telling constraints* in our analysis, and narrow down the number of contracts we have to consider. The truth-telling constraints in our example, under the maximum payment contract, are:

$$E(\Pi_i)^T = R - 170 + \epsilon > R - 170 = E(\Pi_i)^L \ (189 \leq R \leq 200)$$

$E(\Pi_i)^T = R - 0.9R + \epsilon > R - 0.9R = E(\Pi_i)^L$
or
$E(\Pi_i)^T = R - 0.9R + \epsilon > R - 170 = E(\Pi_i)^L$ ($100 \le R < 189$) (4.2)

Constraint (4.2) simply states that the payoff to the entrepreneur, $E(\Pi_i)^T$, from telling the truth and revealing R is greater by the amount ϵ than the payoff from lying, $E(\Pi_i)^L$. We assume that entrepreneurs who lie by saying that their payoff was less than 189 when, in fact, it was 189 or more, are monitored and pay 170, so that lying does not allow them to reduce their payments to the intermediary. Epsilon truthfulness implies that those who receive a payoff of more than 189 do not wish to understate or overstate their payoff, since so doing will cause them to lose the amount ϵ, which they gain in utility from truth-telling rather than lying, and will yield them no financial advantage. Similarly, the payoff for entrepreneurs whose projects yield less than 189 is greater by the amount ϵ if they tell the truth and then be monitored than if they lie and then be monitored; again, there is no financial advantage to be gained by lying, so the entrepreneur will tell the truth. Clearly, no entrepreneur whose project payoff is less than 189 has an incentive to claim a payoff of 189 or more, as this will only increase the amount he would pay to the bank from whom he obtained funding; this explains the second truth-telling constraint for the case where R is less than 189.

Under the critical value contract the payoff to the entrepreneur if he lies about a return greater than 189 and says it was only 189 is R minus 170, which exceeds the payoff of R minus $0.9R$ plus ϵ from telling the truth; this contract does not therefore satisfy the truth-telling constraint that the payoff from telling the truth should be greater than or equal to the payoff from lying.

Notice that the truth-telling constraints hold for the maximum payment contract under the assumption that banks monitor whenever an entrepreneur announces a return of less than some critical value. Clearly, if banks never monitored, the entrepreneur would have an incentive to lie and announce the lowest possible return when, in fact, the return was a higher value.[2]

Bank competition and the standard debt contract

We have shown above that a bank could gain by offering a maximum payment contract rather than committing itself to monitor each project it funds. Entrepreneurs would also gain by accepting the maximum payment contract in preference to the contract with monitoring. Entrepreneurs gain because they keep a greater share of any project return which exceeds 189 than under the contract with monitoring of each project. The maximum payment contract would therefore be attractive to entrepreneurs and offer supernormal profits to a bank faced with competitors committing themselves to monitor all the projects that they fund. Competition would therefore lead all banks to reject the contract with monitoring of each funded project in favour of the maximum payment contract.

Since all banks would be making profits under the new contract, they would compete against each other by offering improved contract terms to try

62   *Investment Finance*

to attract entrepreneurs away from their competitors. This competition between banks would continue until the banks were making zero supernormal profits in a competitive equilibrium and the contract was the standard debt contract.

Notice that the idea of competition between banks tells us not only that in equilibrium banks must be making zero supernormal profits and satisfying the competition constraint, but also that the equilibrium contract must be the most attractive contract to entrepreneurs, consistent with banks satisfying the competition constraint. In other words, whenever the competition constraint is satisfied but the utility of entrepreneurs is not being maximised subject to this constraint, then a bank has the opportunity to move off the zero competition constraint and make profits while at the same time improving the utility of the entrepreneurs. Competition between banks then forces them to further improve the terms offered to entrepreneurs, until they are forced back on to the competition constraint.

To see how competition produces the standard debt contract, consider how the banks could change the terms of the maximum payment contract to try to attract entrepreneurs away from competitor banks. There are two changes a bank could consider. One change would be to reduce the share of the project return taken by the bank. Using equation (4.1) it is possible to see that with this change the bank could also reduce the critical value and maximum payment. The other change would be to hold constant the share of the project return taken by the bank while reducing the critical value and maximum payment below the levels implied by equation (4.1). Both changes involve reducing the maximum payment, but only the first change combines this with a reduction in the share of project returns taken by the bank. Let us distinguish between these two contracts by calling one the 'high share' contract and the other the 'low share' contract, where the latter is the contract with the lower payoff share being taken by banks.

It is clear that in order to make the same expected rate of return under either of these two new contracts, the bank must set the higher critical value and maximum payment for the low share contract, since reducing the bank's share of the payoff, *ceteris paribus*, clearly reduces the bank's expected return and this needs to be offset by increasing the maximum payment (subject to the value being no greater than that given by equation (4.1)) relative to the high share contract. This can be seen easily by examining Figure 4.2.

Figure 4.2 plots payments to the bank, P, on the vertical axis and project returns, R, on the horizontal axis.[3] Two alternative maximum payment contracts are shown, both of which we assume to be capable of yielding normal profits for the bank.

The steeper ray through the origin represents the contract which requires the higher, 90 per cent, share of project payoff to be paid to the bank. The maximum payment for this contract is P_1, which is paid when the project yields the critical payoff value of CV_1; for project payoffs below CV_1, the bank monitors to verify the payoff.

The contract represented by the shallower ray through the origin has a lower share of the payoff going to the bank. If the critical value for this contract was

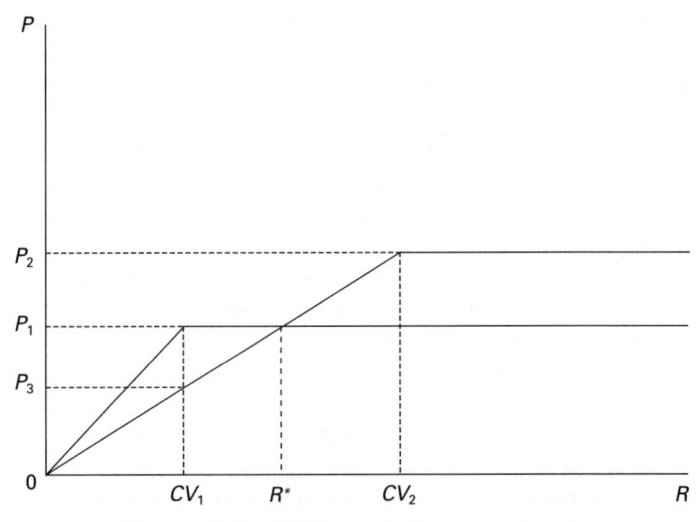

Figure 4.2 Different share contracts

also CV_1, then clearly this project would yield lower returns to the bank than did the other contract; since on all projects yielding a payoff less than CV_1 the bank would receive a lower share, and for other projects it would receive a maximum payoff of P_3 rather than the higher value of P_1, while the bank would spend as much on monitoring under either contract. Thus if each contract is to offer normal expected returns to the bank, then it is necessary to let the critical value for the low share contract rise to CV_2. In this case, the bank will receive more from projects which yield payoffs in excess of R^* under the low share contract than under the alternative, which will offset the fact that it receives less under this contract from projects yielding less than CV_3. It must also be the case, of course, that the extra returns to the bank from projects yielding R^* or more under the low share contract offset the higher expected monitoring costs associated with this contract, which requires monitoring of all projects that yield payoffs of less than CV_2.

The zero profit or competition constraint is satisfied when banks are making an expected percentage return, after monitoring costs, of just enough to cover the rate they must pay to depositors to attract funds, that is, 25 per cent in our example. Thus banks making zero supernormal profits, after monitoring costs, expect to make 25 per cent under either of the two new contracts. In other words, the expected payment by the entrepreneur, $E(P)$, minus expected monitoring costs, $E(MC)$, must equal 125 under either contract. Since the critical value above which monitoring does not take place is lower for the high share contract, it is less likely that monitoring will take place under this contract than under the other contract. Hence expected monitoring costs are lower for the high share contract and, since for both contracts $E(P)$ minus $E(MC)$ equals 125, it follows that $E(P)$ must be lower for the high share contract.

An entrepreneur's expected return from a project net of the expected payment to the bank is given by $E(R)$ minus $E(P)$; that is, the expected project return minus the expected payment to the bank. Since expected project returns, $E(R)$, are constant regardless of the type of contract used in the provision of funds, it is clear that the entrepreneur will prefer the contract with the lower expected payment to the bank. Hence banks offering the high share contract would attract custom away from banks offering the low share contract. Indeed, faced with other banks offering a low share contract, a competitive bank could offer a high share contract which yielded supernormal profits by offering entrepreneurs only a part of the expected savings on monitoring costs. Such supernormal profits would, of course, be competed away by other banks moving to the high share contract and offering even better deals to entrepreneurs to attract their custom.

The logic of the above discussion is that for any two contracts offering the same expected return net of monitoring costs to the bank, the dominant one will be that which has the higher share of project returns going to the bank but the lower critical value and maximum payment. The logical conclusion is, therefore, that competition between banks will drive them to set the maximum value of unity for the share of the payoff which they take and set the lowest possible critical value, beyond which they do not monitor the project payoff, consistent with the zero profit constraint. Interpreting the maximum payment as the repayment of the loan principal, K, plus interest on the loan, rK, shows that the optimal contract may be interpreted as a standard debt contract. Under this contract the bank incurs a monitoring cost whenever an entrepreneur claims to have received a payoff less than the maximum payment and it takes the entire return from the project in such cases.

We have thus shown that the problem of costly state verification leads to the standard debt contract with monitoring of defaulting loans (that is, those for which the borrower declares a payoff less than the required maximum payment) as the optimal response of competitive banks. In our example we have assumed that banks offering depositors a return of 25 per cent can attract sufficient funds to support the projects of all entrepreneurs. The interest rate on loans will be 34.32 per cent, which will yield an average return to banks of 25 per cent once allowance is made for defaulting entrepreneurs and monitoring costs.[4]

Although we have assumed that financial intermediation will take place via banks it is, in fact, easy to see that banks arise naturally in this context. Assume that individual suppliers of funds would not be able single-handedly to fund a project. Hence several suppliers would need to band together to fund a project, which would mean that in the event of default they would each need to monitor, assuming the return to monitoring to be private information. Since a bank needs to monitor each defaulting project only once, this involves a saving on monitoring costs, which explains why banks represent the optimal form of financial intermediation. If the suppliers of funds are risk-averse then there is a further reason for the development of banks; banks invest in many projects and allow small suppliers of funds to diversify and spread risk rather than place all their eggs in one basket.

The possibility of rationing

In the above example the problem of costly state verification does not lead to rationing since we assumed that banks face a perfectly elastic supply of deposits at a deposit rate of 25 per cent. As in previous chapters, it is easy to introduce rationing into the analysis by assuming, instead, that the supply of deposits depends positively on the deposit rate.

A little calculus allows us to show that, for the uniform distribution of project returns specified in the example, the maximum return net of monitoring costs is made by banks charging an interest rate of 90 per cent.[5] Thus there is a critical value for the interest rate that banks will charge, which is dependent upon the probability distribution from which project returns are drawn. Raising the interest rate above this critical value serves only to reduce the net returns to banks and reduce the interest rate they can pay to depositors. If the supply of deposits to banks depends upon the interest rate paid to depositors, then the supply of loans curve becomes backward-bending, as shown in Figure 4.3.

The intuition for the backward-bending loan supply schedule is straightforward. As banks increase the quoted loan rate there are two effects. One is to increase the payments entrepreneurs acquiring loans will expect to make to the bank from which they borrowed; this effect increases the expected returns of the banks. The other effect is that, as the quoted loan rate rises, more entrepreneurs are unable to repay their loans and the banks therefore incur more monitoring costs; this effect reduces the expected returns of banks. For low quoted loan rates the first effect dominates the second, so that as the rate is increased, banks increase expected returns (net of monitoring costs) from making loans and so are able to offer a higher rate to depositors and attract more deposits. However, beyond a certain critical value for the quoted loan rate, the second effect begins to dominate so that increasing the quoted loan rate serves to reduce the expected returns (net of monitoring costs) to banks and to reduce the rate they can offer depositors and the amount of funds they can attract as deposits. If banks increase the quoted loan rate beyond this critical value they attract fewer deposits and are able to make fewer loans. Thus the loan supply schedule is backward-bending and banks will not charge a quoted loan rate in excess of that rate at which the schedule begins to bend backwards.[6]

Figure 4.3 shows two cases. The demand curve in both cases is the kinked function shown, since the entrepreneurs' participation constraints are satisfied for any interest rate below that which makes the maximum payment equal to the highest possible project payoff. Thus for any interest rate below that which makes $(1 + r)K$ equal to b (or $r = $ to $b/K - 1$) all entrepreneurs apply for funds.

Figure 4.3(i) shows the case where rationing does not occur, since the supply and demand curves intersect at point A and an interest rate on loans of r_A, which is below the rate at which the supply curve begins to bend backwards. As drawn, the supply and demand curves also intersect at a rate of r_B in Figure 4.3(i), but this rate will not be charged since, if it were, competition between banks would drive the rate down to r_A, as the reader should be able to show.

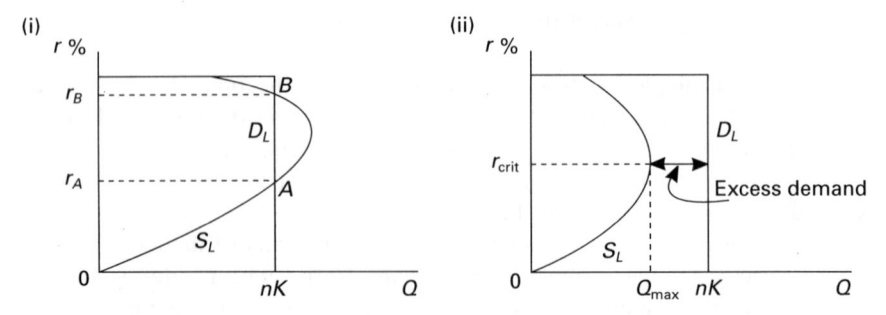

Figure 4.3 Hidden information and the credit market

Figure 4.3(ii) shows the rationing case. Here, the supply and demand curves intersect only along the horizontal part of the demand curve. Since banks have no incentive to raise the interest rate above the level of r_{crit}, at which the supply curve begins to turn round, the market interest rate will be r_{crit} and there will be an excess demand for loans as shown, with demand equal to nK and supply equal to Q_{max}. Loan applicants will be randomly rationed and unsuccessful applicants will be unable to bid loans away from other entrepreneurs or draw additional funds into the market, since banks would lower their net returns if they charged an interest rate above r_{crit}.

Perhaps unexpectedly, the problem with rationing in a case like that shown in Figure 4.3(ii) is not that too few funds are provided for investment. On the contrary, it is possible to show that too much investment is funded! The logic behind this result is that monitoring is costly and a reduction in either the deposit rate (or, equivalently, the quoted loan rate, since this would then lower the deposit rate) which led to fewer projects being funded, would also lead to a saving on expected monitoring costs, both because fewer projects would be funded and because those funded would be less likely to default at a lower interest rate. Over a certain range it is possible to show that this reduction in monitoring costs would be greater than the loss due to funding fewer projects, so that it would be beneficial, in terms of the net gains to society as a whole, to invest in fewer projects rather than more.

The gains from reducing the interest rate and investment are not spread evenly over all members of society: those entrepreneurs fortunate enough to obtain funding at the reduced interest rate would gain, but at the expense of other entrepreneurs now denied funding and of the suppliers of funds who now gain a lower return on deposits with banks. Nevertheless, it is possible to use this argument in favour of capping the interest rate banks may pay to depositors. Such a policy would have the effect of reducing the funds made available for investment. It is possible, but not necessarily so, that this would increase the gains to society as a whole, even in the case where the market would not, of its own, lead to rationing. For this to be the case, all that is needed is that the net gains to society from funding the marginal investment project are less than the potential savings on monitoring costs available by reducing the deposit rate of interest and the amount of investment.

4.3 The credit market and business cycles

In this section we trace the outlines of a theory of business cycles based on a problem of costly state verification. Assume that an economy in any one period consists of two generations of people. The younger generation are workers. Workers have an endowment of labour which they sell on the competitive labour market to entrepreneurs. Workers save their wages and become entrepreneurs in the following period. The older generation act as risk-neutral entrepreneurs, each of whom has some wealth, w, and is associated with an investment project. The entrepreneurs may invest their wealth in a storage technology, which yields a certain rate of return of ρ, or in a bank, or in their own investment project, which may require funding greater than w if it is to be carried out. The value of w is common to all entrepreneurs and is, in fact, just their savings from working in the previous period for the previous generation of entrepreneurs. Entrepreneurs consume all their returns at the end of each period and then die. Thus individuals live for two periods only. A new generation of workers is born each period and the population of the economy remains constant.

The returns to investing in projects are random. All projects are alike in that the return to each will be given by an independent drawing from the same distribution as every other project, so that all have the same expected return (net of labour costs), $E(R)$.[7] Projects are, however, different in an important way since the investment cost of a project, C_i, differs across entrepreneurs. The idea here is simply that some entrepreneurs are better than others and able to invest more cheaply.[8]

The full information case

Consider, initially, a world of symmetric information where everyone can observe freely the cost associated with any project and also the return produced by any project. In this case, an entrepreneur would wish to invest in his project if it yielded him an expected return of ρ^* or greater, and he would be able to obtain funding for it if it cost more than his initial wealth of w so long as he could pay to the suppliers of funds an expected rate of return equal to ρ^*, since nobody will invest in a project that offers a rate of return less than ρ^*, which can be achieved by investing in the storage technology.

In this case, whether investment funds were supplied via an equity or a credit market, all projects capable of yielding an expected rate of return of ρ^* or more would be able to acquire funding. These would be the cheaper projects. Let the project which costs C^* be capable of producing an expected rate of return of ρ^*,

68   Investment Finance

then projects costing C^* or less will obtain funding, since they offer a return greater than or equal to ρ^*, while those costing more than C^* will not be funded.

This is illustrated in Figure 4.4, which plots the expected excess return to an entrepreneur from his project, X, on the vertical axis, against the investment cost, C, on the horizontal axis. The expected excess return is seen easily to be given by:

$$X = E(R) - w(1 + \rho^*) - (C - w)(1 + \rho^*) \tag{4.3}$$

Equation (4.3) holds, since the entrepreneur must subtract from the project's expected return of $E(R)$ an amount of $w(1 + \rho^*)$ to cover the opportunity cost of his own funds invested in his project, which could otherwise have been invested at the safe rate of return from storage, and an amount of $(C - w)(1 + \rho^*)$ to cover the expected payments he must make to the suppliers of funds. Note: this does not mean that the interest rate on loans will be ρ^*; rather, for projects with a possibility of default the interest rate will exceed ρ^* to allow for the fact that sometimes the project will fail; the expected rate received by lenders is, however, ρ^*, and on safe projects with no default risk this will, in fact, be the interest rate charged on loans.

There will be no risk of default for projects which are so cheap that they can be funded either from the entrepreneur's own wealth or with sufficiently small borrowing that even the lowest possible project return could yield a return of ρ^* on the debt used to finance it. The interest rate charged on loans will depend on the probability of default and will be higher for costlier projects than for cheaper ones.

The curved line, FI, in Figure 4.4 shows that, since the expected return, $E(R)$, is common across projects, the expected excess return, X, falls as C rises. The intersection of the line FI with the horizontal axis at C^* shows that projects with a cost less than or equal to C^* will be funded since they yield their entrepreneurs a positive expected excess return compared to investment in the storage technology or in banks. Having determined C^* we then, from a knowledge of the distribution of projects across entrepreneurs, could determine how much investment would take place in the economy; let the amount be denoted Q^*.

Notice from equation (4.3) that the terms in w may be cancelled to yield:

$$X = E(R) - C(1 + \rho^*) \tag{4.4}$$

Equation (4.4) shows that the marginal value for C, which yields X equal to zero, is independent of the wealth level of entrepreneurs. Hence, under full information, the amount of investment of Q^* which will be carried out is independent of the wealth level of entrepreneurs, provided only that the aggregate level of wealth is sufficient to fund that level of investment.

The asymmetric information case

Now consider that there is an asymmetry of information in the economy. Only the entrepreneur associated with a project is able to observe freely the return that it yields; anyone else wishing to observe this return must pay a monitoring cost of c. This type of asymmetry of information, as we know from the previous

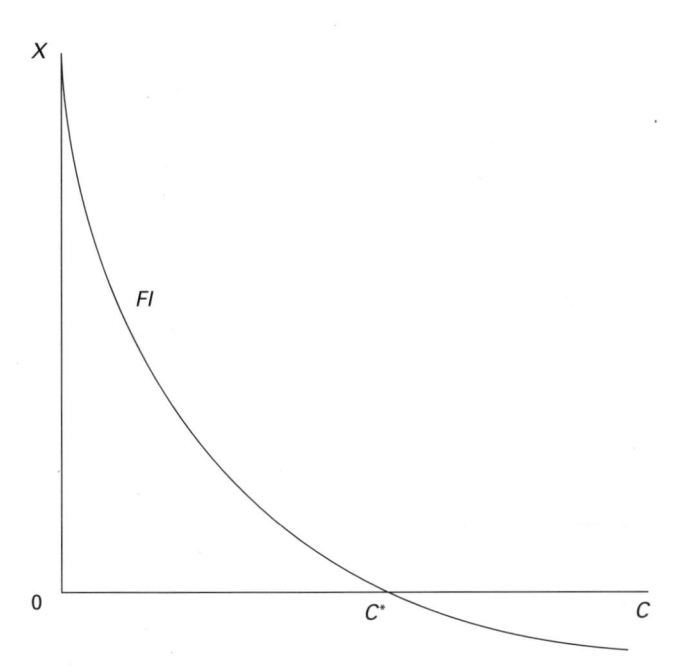

Figure 4.4   Investment cost and expected excess returns under full information

section, leads to finance being provided via a credit market and banks. It can also be shown that it is likely to lead to a reduction in investment compared to the full information case and an important role for the entrepreneur's own wealth, w. The intuition for these latter results is relatively straightforward and may be illustrated using Figure 4.5.

Figure 4.5, like Figure 4.4, shows the relationship between X and C. The *FI* line in Figure 4.5 is exactly the same as that in Figure 4.4 to enable comparisons to be drawn easily. The figure shows two *CSV* lines, one drawn for the costly state verification case, when entrepreneurs have a wealth level of w_1, and one for the wealth level of w_2.

Consider the line $CSV(w_1)$. This line shows the excess expected return which projects could offer to the entrepreneur, X, against investment cost, C. The value of X must now be determined as follows:

$$X = E(R) - w(1 - \rho^*) - (C - w)[1 + \rho^* + E(MC)] \tag{4.5}$$

Equation (4.5) differs from equation (4.4) because of the extra term $E(MC)$. This term is introduced to allow for the fact that defaulting projects will now be monitored and so entrepreneurs have to pay an interest rate to lenders which allows them to cover not only the cost of funds of ρ^* but also the expected monitoring costs (expressed here as a percentage premium on the rate lenders expect to pay).

For very cheap projects with a cost of less than C_A, the line $CSV(w_1)$ is the same as the line *FI*. Projects costing less than C_A are so cheap that either the

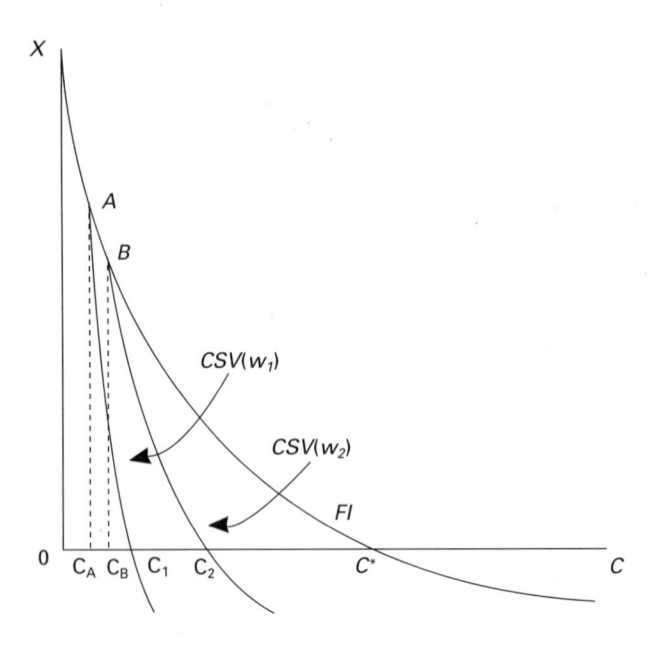

Figure 4.5 The impact of costly state verification

entrepreneur's wealth of w_1 is sufficient to fund the project, so that no credit market funds are needed and all the project's expected return goes to the entrepreneur, or that, although the entrepreneur needs to borrow to fund his project, the minimum possible project return is capable of providing a return of ρ^* to lenders. Lenders are therefore guaranteed a return of ρ^* if they invest in such projects and will never need to monitor the project payoff. There is, therefore, no problem of costly monitoring associated with these projects; the *CSV* and *FI* lines are coincident to the left of point A, since for such projects $E(MC)$ is zero. Such projects will be able to attract funds at the safe rate of interest of ρ^* exactly as they would in a full information world.

Projects which cost more than C_A, however, are unable to guarantee a return of ρ^* to lenders. For such projects, lenders will have to charge an interest rate above ρ^* in order to achieve an expected return of ρ^* since sometimes the project will default and be monitored. Hence for the costly state verification case (and entrepreneur wealth of w_1) there is a possibility of monitoring costs associated with projects with a cost in excess of C_A. These expected monitoring costs cause the *CSV* line to diverge from the *FI* line as the project investment cost rises and, since monitoring is more likely the more costly the project, this divergence increases with investment cost.

Thus, under costly state verification, the marginal project capable of offering its entrepreneur an expected return of ρ^*, or, equivalently, a zero value for X, is that which costs C_1 rather than C^* as for the full information case.

Projects costing between C_1 and C^* will no longer be carried out. The reason for this is simple. Although the expected gross rate of return on such projects exceeds ρ^* (and this allows the entrepreneur to offer lenders an expected rate of return of ρ^* while achieving a better rate than that for himself in a full information world), once the asymmetry of information and expected monitoring costs are introduced the effect is to reduce the excess return, X, for any project which has a positive probability of default. Projects costing more than C_1 offer their entrepreneurs a negative excess return once allowance is made for expected monitoring costs, and so entrepreneurs no longer wish to seek funding for them.

The line $CSV(w_2)$ is drawn for a higher level of wealth held by each entrepreneur. The higher wealth level has the effect that entrepreneurs need to borrow less to fund their projects. This, in turn, has two effects. First, more projects are able to offer a guaranteed rate of return of ρ^* to lenders, thus extending the section over which the CSV and FI lines are coincident to point B. Second, the cost of the marginal project capable of offering its entrepreneur a zero excess return is increased to C_2. This latter effect occurs because, as entrepreneurs borrow less to fund their projects, it becomes less likely that they default for any given interest rate, thus reducing expected monitoring costs and hence the interest rates that lenders charge to entrepreneurs. The reduction in the interest rates charged to entrepreneurs for the higher wealth level causes the marginal project to have a cost of C_2 which is greater than C_1.

Since the marginal project for which an entrepreneur wishes to acquire funding varies with the level of wealth held by entrepreneurs it follows that, in the asymmetric information world, the aggregate level of investment, Q, depends upon the level of wealth held by entrepreneurs; Q will, in general, be less than the value of Q^* which applies for the full information world. Although for sufficiently high levels of w the two values will coincide, it will never be the case that Q will exceed Q^*, since investment in projects beyond Q^*, or in projects costing more than C^*, mean obtaining an expected return of less than that available from investment in the safe storage technology even in the absence of monitoring costs. The relationship between Q and w is plotted as the line $Q = q(w)$ in Figure 4.6.

The relationship $Q = q(w)$ slopes upwards, showing that Q rises as w rises until w reaches a value of w^* and Q reaches Q^*, which is the full information level of Q. Hence for values of w greater than or equal to w^* there is no risk of default on the marginal project, which has an expected gross rate of return of ρ^* and the aggregate level of investment will be equal to that in the full information world.

Assume that the higher the level of investment in an economy the higher will be the productivity of labour. Assuming also that labour is paid a wage equal to its marginal product, it then follows that w rises as Q rises. This latter relationship is shown as the line $w = h(Q)$ in Figure 4.6. The intersection of the two lines in the figure at w_E and Q_E represents the equilibrium values for w and Q.

It is possible that the intersection of the two lines in Figure 4.6 could produce the same level of aggregate investment as the full information world, but as we

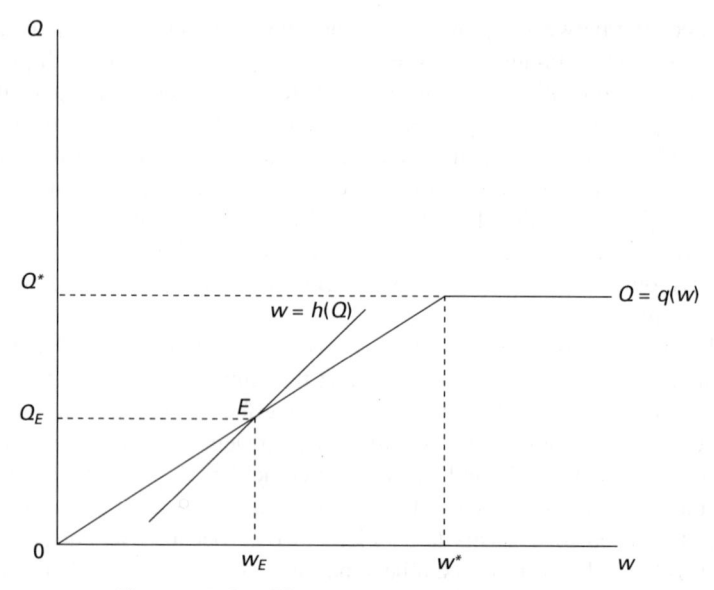

Figure 4.6 Macroeconomic equilibrium

have drawn it Q_E is less than Q^*. The result of the asymmetry of information may, therefore, be to reduce both aggregate investment and the level of labour productivity.

Cyclical effects may be introduced into this model if we assume that the productivity of investment, and hence the productivity of labour, is subject to random shocks; sometimes productivity will be above average and sometimes below average. The productivity shock is not revealed, however, until the investment has been carried out. Therefore, entrepreneurs and the suppliers of funds base their decisions on the expected value for the productivity of investment which is a constant. The relationship $Q = q(w)$ therefore will not be affected by introducing shocks into the model.

Note that there will be no effect upon the level of aggregate investment produced in a full information world, where the aggregate level of investment is determined only by the constant expected productivity of investment. The impact of the productivity shocks in a full information world will therefore only be to produce random shocks to wages and, hence, savings. The impact is much more significant in the case of asymmetric information, as Figure 4.7 shows.

Figure 4.7 repeats Figure 4.6, but shows two different relationships between w and Q. The relationship $w = h(Q)$ represents the case when the productivity of capital is at its average value, while the relationship $w = h'(Q)$ represents the case where there has been a positive shock to the productivity of capital. Imagine that the economy is initially in equilibrium at Q_E and w_E when there is a positive shock which lasts for one period only, after which productivity again returns to its average level with no further shocks. This one-off shock to produc-

The Costly State Verification Problem 73

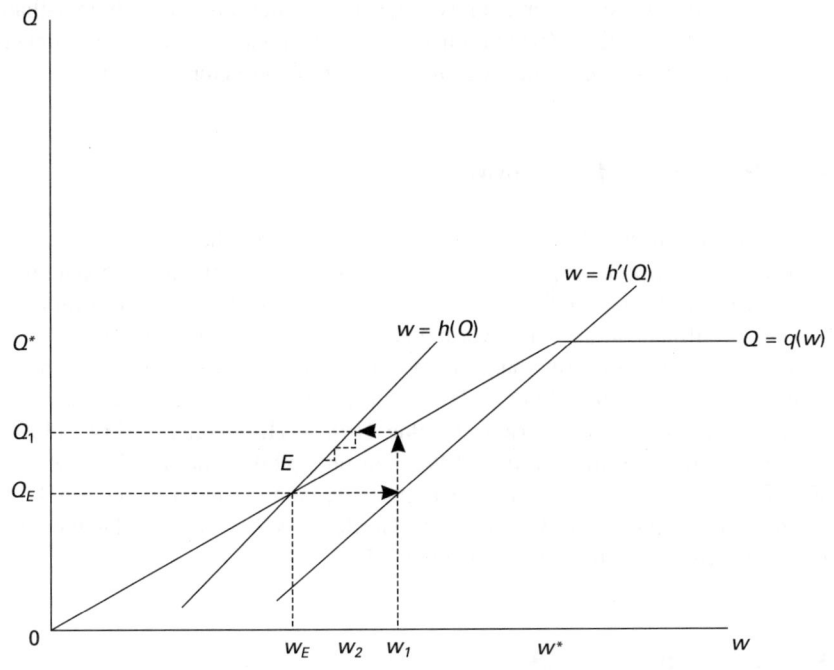

Figure 4.7 Persistence effects of shocks

tivity will have a long-lived effect on the economy and the path taken will be like that shown in the diagram.

The dynamic impact of the productivity shock is easily explained. The aggregate investment level when the shock occurs is Q_E so that, given the temporarily high level of productivity, wages will be given by w_1, which is higher than the usual value of w_E. The higher level of wages produces higher savings and a higher level of capital stock of Q_1 in the next year, which is found by reading off from the $Q = q(w)$ relationship. Even though the productivity of capital has, by now, returned to its average level, the higher-than-average level of aggregate investment maintains wages above their average level – they are given by w_2 on the $w = h(Q)$ line. These higher-than-average wages produce higher-than-average savings, which produce higher-than-average aggregate investment, which produces higher-than-average wages and so on until eventually, assuming stability, the economy returns to the equilibrium point E as shown.

Thus, although productivity shocks in the full information version of our model produce only temporary effects, in the asymmetric information case they lead to long-lived effects or persistence of the sort observed in real business cycles, in which it is typical for higher-than-average output to be followed by higher-than-average output (and vice versa for lower-than-average output). This type of effect may have some role to play in explaining cyclical behaviour in the real world and would appear to be consistent with the empirical evidence that when firms have higher-than-average profits they also have higher-than-average retained earnings (which mirrors w in our model) and carry out higher-than-average investment (see, for example, Fazzari and Petersen, 1993).

4.4 Recommended reading

An excellent, but possibly difficult, paper showing how the standard debt contract is the optimal response to the costly state verification problem is Williamson (1986), which also shows the possibility of credit rationing. Although Williamson assumes that defaulting projects are monitored with certainty, it is possible to show that randomised monitoring is the optimal response to the costly state verification problem and the reader interested in this development is referred to Mookherjee and Png (1989). Hillier and Worrall (1994) examine the welfare implications of Williamson's model, and develop the argument for capping the interest rate. The macroeconomic model of section 4.3 is based upon the paper by Bernanke and Gertler (1989) and a similar model is also to be found in Hillier and Worrall (1995).

4.5 Problems

Problem 4.1
Assume that there are n entrepreneurs, each of whom is endowed with one project. Project returns, R, are drawn from a uniform distribution with a equal to 100 and b equal to 150. Projects cost 100 and entrepreneurs have no funds of their own. There is a problem of costly state verification and the monitoring cost is 5.

(a) What is the project expected gross return, $E(R)$?
(b) Will any projects be funded if banks must pay 25 per cent to attract deposits? Explain your answer.
(c) If banks must pay 10 per cent to attract deposits and monitor all projects that they fund, what percentage of project returns, R, would banks need to take in order to break even?
(d) Would the share contract with monitoring of part (c) above produce a Nash equilibrium? Explain your answer.

Problem 4.2
Show how credit rationing may be explained by models of asymmetric information. What are the strengths and weaknesses of the models you discuss? (Your answer should make use of arguments drawn from the first four chapters of this book and not just from the present chapter.)

Part II
Asymmetric Information Problems in the Insurance Market

CHAPTER 5

Insurance and Risk Aversion

5.1 Overview

In the previous chapters we assumed that all individuals or organisations were risk neutral, that is they were concerned only with the expected yield from an uncertain situation. For many purposes, however, the assumption of risk neutrality may be inappropriate and it is necessary to consider alternative assumptions. In section 5.2 we review the alternative attitudes which may be taken towards risk, and in section 5.3 we show why an individual may wish to purchase insurance to offset risk in a symmetric information setting. The following chapters show how asymmetric information affects the market for insurance: Chapter 6 examines the effects of moral hazard with hidden action, and Chapter 7 looks at the effects of adverse selection. The present chapter closes with some recommended reading and two problems.

5.2 Attitudes towards risk

We define a *risky situation* or *prospect* as one which has associated with it a set of outcomes or payoffs, where each outcome is associated with a particular state of the world which occurs with some positive probability. The sum of the respective state-probabilities is unity, since one of the states must occur. For example, a lottery ticket with a 50 per cent chance of a payoff of £10 and a 50 per cent chance of a payoff of zero represents just such a risky prospect, with the two states of the world being win or lose.

An individual's attitude towards risk may be categorised at a very simple level by asking whether he would prefer a certain payoff of £5 to the lottery ticket described above. The expected return from the lottery ticket is easily seen to be £5, so we say that an individual who is indifferent between the certain payoff of £5 and the lottery ticket, with its expected return of £5, is *risk neutral*.

78   The Insurance Market

An individual who would prefer to be given the lottery ticket rather than to be given £5 is said to be *risk loving*, while an individual is said to be *risk averse* if he would prefer the certain payoff of £5 to the lottery ticket.

If we make the assumption that individuals calculate the expected utility of a risky prospect by evaluating the mathematical expectation of the utilities associated with each of the possible project payoffs, there is a straightforward relationship between attitudes towards risk and the marginal utility of wealth. Consider the simple lottery once more. Figure 5.1 shows the utility of wealth for an individual. The diagram has been normalised so that the value of zero at the origin represents the individual's existing wealth level rather than true zero.

Consider first the straight-line utility function in Figure 5.1(i); this shows an individual with a constant marginal utility of wealth; that is, the second derivative of the utility function, $U''(W)$, equals zero.[1] If this individual was given the lottery ticket his expected wealth would be £5, which we indicate on the diagram by the notation of $E(R)$ below the £5 point on the horizontal axis.

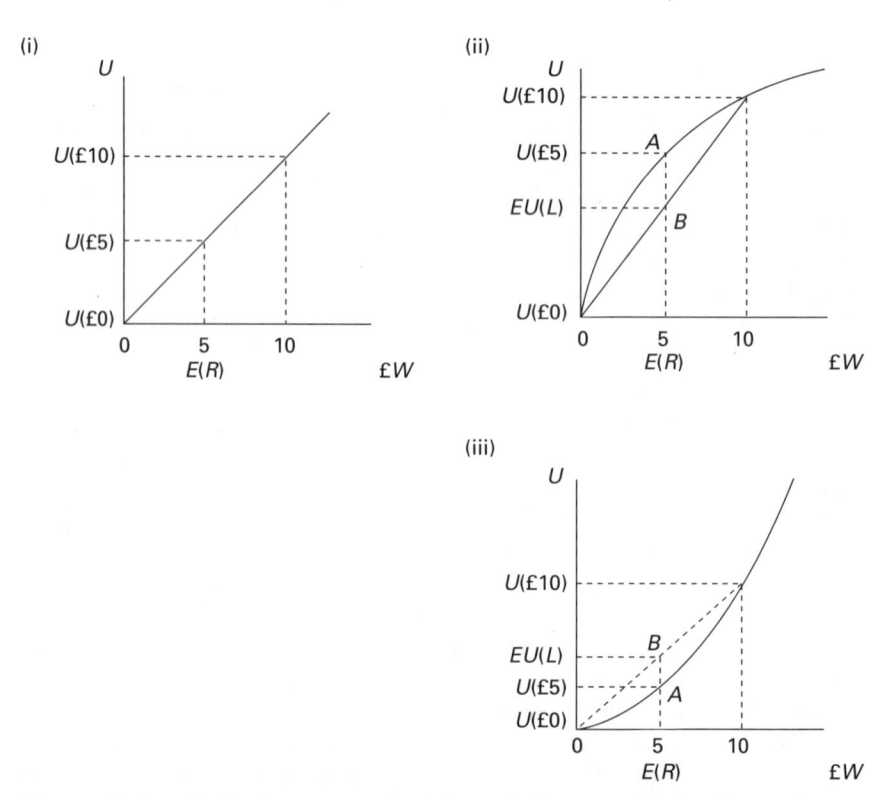

Figure 5.1   Attitudes towards risk and the marginal utility of wealth

Calculating this individual's expected utility from ownership of the lottery ticket according to our assumption above yields:

$$EU(L) = 0.5U(£0) + 0.5U(£10) \qquad (5.1)$$

where $EU(L)$ is the expected utility from owning the lottery ticket and this lies half-way between the utility level to be gained from zero wealth and the utility level to be gained from £10. Inspection of the diagram makes it clear that $EU(L)$ is exactly equal to $U(£5)$, the level of utility to be gained from a certain payoff of £5.

Thus, for this individual, the expected utility from owning the lottery ticket is equal to the utility to be gained from a certain wealth equal to the expected return from the lottery ticket; that is, for this individual, $EU(L)$ equals $U(£5)$ equals $U[E(R)]$, where $U[E(R)]$ represents the utility to be gained from a certain payoff equal to the expected payoff from the lottery ticket. This individual may therefore be said to be indifferent between £5 for certain or the lottery ticket with an expected return of £5. Thus under our method of calculating expected utility, risk neutrality is associated with constant marginal utility of wealth.

Now consider the concave utility function, which shows diminishing marginal utility of wealth, in Figure 5.1(ii). In this case, the second derivative of the utility function, $U''(W)$, is less than zero. Once more, the expected return from the lottery ticket is £5. If we read up from the £5 value on the horizontal axis to the curved line at point A, we can read across to the vertical axis to see the utility level associated with a payoff of £5 for certain, which we denote $U(£5)$ which equals $U[E(R)]$. The expected utility to be gained from holding the lottery ticket is again given by equation (5.1) and, as before, is located vertically halfway between $U(£0)$ and $U(£10)$. $EU(L)$ may be found by reading up from $E(R)$ on the horizontal axis until the intersection at point B with the chord formed by joining the points on the utility function for wealth levels of £0 and £10. Reading across from B to the vertical axis shows $EU(L)$. This time, however, the resulting value for $EU(L)$ does not coincide with the value for $U[E(R)]$ or $U(£5)$, but lies below it. This individual would therefore prefer a payoff of £5 with certainty to a lottery ticket with an expected payoff of £5; he is therefore said to be risk averse, since he dislikes risk. Thus individuals with diminishing marginal utility of wealth are risk averse under our method of calculating expected utility.

This diagrammatic technique for comparing $U[E(R)]$ and $EU(L)$ works for any value of p, since reading up from $E(R)$ on the horizontal axis to the straight line chord attaches the same probability weights to utilities in the vertical plane as are attached to wealth levels in the horizontal plane. When the utility function is a straight line, as in Figure 5.1(i), points A and B are coincident, but for curvilinear utility functions, as in Figure 5.1(ii) and (iii), points A and B differ.

Now consider the convex utility function, which shows increasing marginal utility of wealth, in Figure 5.1(iii). In this case, the second derivative of the utility function, $U''(W)$, exceeds zero. Once more the expected return from the lottery ticket is £5. If we read up from the £5 value on the horizontal axis to the curved line at point A we can read across to the vertical axis to see the utility level associated with a payoff of £5 for certain, $U(£5)$ or $U[E(R)]$. As above, it is possible to show $EU(L)$ by reading up from $E(R)$ on the horizontal axis until the intersection at point B with the chord formed by joining the points on the utility function for wealth levels of £0 and £10. Reading across from B to the vertical axis shows $EU(L)$, which is clearly greater than $U(£5)$ or $U[E(R)]$. This individual would therefore prefer the lottery ticket to a payoff of £5 with certainty; he is, therefore, said to be risk loving, since he likes risk. Thus individuals with increasing marginal utility of wealth are risk loving under our method of calculating expected utility.

5.3 Risk aversion and insurance

For the remainder of this chapter we shall assume that individuals are risk averse and show that this leads them to demand insurance against risks. Since most individuals do buy insurance against large risks, such as risks to their car or house, it seems reasonable to assume that most people are risk averse. The reader may, however, wonder why it is that many people, even those who insure their property, are observed to gamble by buying lottery tickets, visiting casinos or placing bets, which would seem to indicate that they are not risk averse but are, instead, risk lovers. The best explanation of this puzzle seems to be that people find the act of gambling to be fun; they are therefore prepared to pay small amounts for the pleasure of watching the lottery numbers being drawn on television and feeling that they can win something, or enjoy watching a race on which they have placed a bet. Thus, for most people, gambling should be viewed as a recreational activity involving repeated small costs. However, when it comes to large amounts and serious risks, such as having a house burn down or property stolen, behaviour is better characterised by assuming risk aversion than risk loving.

The demand for insurance

Consider a risk averse individual who owns an expensive item of property, his car, say, which is at risk of being stolen. Let the individual's total wealth be T, let the car be worth the amount C, and let the probability of theft be p. Assume that the probability of theft, p, is given exogenously: that is, it does not depend upon the actions of the individual, and that the value of p is known to both the individual and to the insurance company offering insurance. These assumptions rule out some interesting problems of asymmetric information which we will introduce in the remaining chapters of this section.

The problem facing this individual may be illustrated using Figure 5.2, which is similar to Figure 5.1(ii) above. Figure 5.2 shows a risk averse individual's utility of wealth. Consider the situation facing this individual if he *does not*

Insurance and Risk Aversion 81

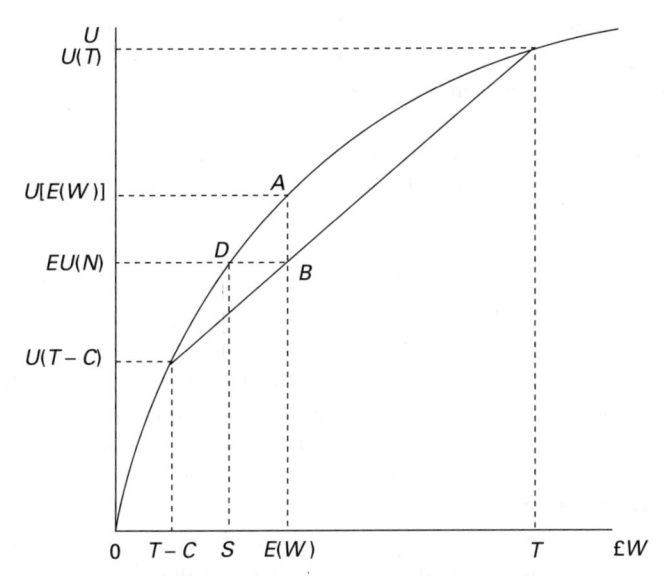

Figure 5.2 Risk aversion and insurance

insure. If his car *is not stolen*, his wealth will be T and his utility will be $U(T)$, while if his car *is stolen*, his wealth will be $(T - C)$ and his utility will be $U(T - C)$ as shown on the diagram. His expected wealth, $E(W)$, is given by:

$$E(W) = p(T - C) + (1 - p)T \tag{5.2}$$

and his expected utility, $EU(N)$, where we use N to indicate that he has no insurance, is given by:

$$EU(N) = pU(T - C) + (1 - p)U(T) \tag{5.3}$$

Since this individual is risk averse, his expected utility from the risky situation, $EU(N)$, is less than the utility, $U[E(W)]$, he would get from a certain level of wealth equal to the expected wealth of $E(W)$; this is shown on the diagram by the point B being vertically below the point A.

Notice that a level of wealth of S for certain would yield the same level of utility as $EU(N)$, as is shown by reading along from B to D in the diagram and then reading down to the horizontal axis at S. Thus, if an insurance company promised to replace the individual's car, or pay him the amount of *compensation*, Y, equal to C, if it was stolen, he would be willing to pay a *premium* of X for this *insurance contract* as long as X was no greater than $(T - S)$, since this would then guarantee him an income level of S or more.

When the compensation paid by the insurance company to the customer, or insuree, in the event of the hazard occurring (that is, the theft of the car in this case) is enough to fully compensate the customer (so that the event of the hazard occurring does not make the customer feel any worse off), the insurance company, or *insurer*, is said to provide *full insurance*. Initially, we shall consider only such full insurance, although we shall deal with more general contracts later.

Once fully insured, the individual becomes certain of his wealth level; if no theft occurs, his wealth level is $(T - X)$, and if theft occurs it is still $(T - X)$ since the loss of the value of his car is made good by the insurance company. As long as the premium X is less than $(T - S)$, the individual ensures a certain level of wealth of more than S by paying the premium to purchase the insurance. Since $U(S)$ is equal to $EU(N)$, with an insured level of wealth greater than S he obtains more utility than his expected utility in the uninsured state and is therefore willing to purchase insurance. If the insurance premium was more than $(T - S)$, for the compensation amount C, the individual would prefer to remain uninsured as his expected utility level without insurance would then be in excess of the certain utility level he would obtain if insured.

An individual purchasing full insurance is paying the premium, X, in return for an uncertain repayment from the insurance company; the actual repayment will be zero with probability $(1 - p)$ (that is, when the car is not stolen), and C with probability p (that is, when the car is stolen). In a sense this is taking a gamble, since he makes a certain payment in return for an uncertain return. To avoid confusion we say that an individual is purchasing insurance when the risks he is accepting offset other risks to which he is exposed; for example, the compensation from insurance is received in the event of the car being stolen.

Notice that it is wise to avoid loose phrases such as, 'A risk averse person is one who would refuse a gamble and would purchase insurance.' This phrase is inappropriate since we know that he will purchase insurance only if it is not too costly and, conversely, would, unless extremely risk averse, accept a very favourable gamble such as a lottery ticket costing £1 with a 50 per cent chance of a return of £1 million and a 50 per cent chance of a return of zero. A better description of a risk averse individual is, therefore, as follows: 'A risk averse person is one who prefers a level of wealth of Z with certainty to a risky prospect with a level of expected wealth equal to Z.' The reader ought to work out the equivalent descriptions for risk neutral and risk loving individuals.

The supply of insurance

We know that a risk averse person will insure his car against theft as long as the insurance premium, X, for a compensation payment of C, is less than $(T - S)$. Now consider the premium that an insurance company would require if it was to offer insurance to the individual. We assume that insurance companies are risk neutral and that the market for insurance is competitive.[2] Thus insurance companies compete against each other for customers by reducing the premium they charge to a level that yields them only normal profits, which, as in previous chapters, we assume to be zero for convenience. We also assume for convenience that the costs of running the insurance company are zero. Hence, in equilibrium, assuming full insurance, the premium, X, is given by

$$X = pC \tag{5.4}$$

Equation (5.4) is easily interpreted. The insurance company receives the premium, X, with certainty and pays the compensation, Y, which we for now assume equals C, with probability p; thus its expected outgoings equal its receipts and expected profits are zero as long as Equation (5.4) holds and the premium is exactly matched by the expected compensation payment. Insurance at this cost is said to be insurance at *fair odds*; if X exceeds pC it is said to be at *unfair odds*; and if X is less than pC it is said to be at *favourable odds*. Equation (5.4) could therefore be termed the zero profit or competition constraint for this example, or it could be termed the fair odds constraint.

Similarly, a gamble is said to be fair if the expected return net of the cost of partaking in it is zero; for example, the simple lottery ticket discussed in section 5.2 would represent a fair gamble at a cost of £5 for the ticket equal to the expected return from the lottery of £5, thus giving an expected net return of zero to the purchaser of the ticket (the seller also has a zero expected net return). An unfair gamble has a negative expected net return, and a favourable gamble has a positive expected net return (that is, the purchaser of an unfair lottery ticket would have a negative expected return and the seller would have a positive one, and vice versa for a favourable lottery ticket).

The certain wealth level of the individual after purchasing full insurance at the premium of X equal to pC is $T - pC$, which examination of equation (5.2) and a little manipulation shows to be equal to $E(W)$, his expected wealth in the uninsured state. Since $E(W)$ is greater than S, the certain level of wealth which yields the insuree a level of utility equal to his expected utility if uninsured, it is clear that the insuree gains an increase in utility by purchasing full insurance at fair odds; that is, his utility after purchasing full insurance, $UE(W)$, exceeds his expected utility if uninsured, $EU(N)$. In other words, the premium pC he pays is less than the amount $(T - S)$, which is the amount he could pay and remain indifferent between purchasing full insurance or not.[3]

Full insurance at fair odds

We have assumed so far that the insuree receives a compensation level, Y, equal to C, the value of his car. In general, however, it is possible for Y to be more or less than C. Regardless of whether Y is greater than, less than or equal to C, insurance is still said to be fair as long as the premium, X, equals the expected value of the compensation, pY; that is, the value of the compensation times the probability of receiving it. Here p may be interpreted as the price of a unit of compensation, Y as the amount of compensation purchased and X as the total cost of purchasing insurance providing compensation of Y. Insurance is offered at fair odds if the price of a unit of compensation is equal to p, the probability of theft, and is offered at unfair odds if the price is greater than p, or at favourable odds if the price is less than p.

It is easy to see that the insuree would prefer full insurance at fair odds to any other level of insurance at those odds, that is Y equal to C rather than any other value. To see this, consider Figure 5.3. The figure is similar to Figure 5.2, in that the level of utility under full insurance is $UE(W)$, as shown by reading up from $E(W)$ to the point A and across to the vertical axis. $UE(W)$ is, as before, greater than the level of expected utility in the absence of insurance, $EU(N)$, as shown by point B.

Now consider the case of *partial insurance*, that is the case where Y is less than C. Under partial insurance the level of wealth of the individual if no theft occurs is given by $(T - pY)$ and his level of wealth if theft occurs is $(T - pY - C + Y)$. Thus the level of wealth in the no-theft state is decreased, for example to H as shown in Figure 5.3, and the level of wealth in the theft state is increased, for example to I in Figure 5.3, compared to remaining uninsured, but these two wealth levels are not made equal, at $E(W)$, as under full insurance.

Partial insurance at fair odds does not affect the expected level of wealth, which remains at $E(W)$.[4] The expected utility level under partial insurance is found, therefore, by reading up from $E(W)$ on the horizontal axis to point J on the chord drawn between the points K and L on the utility function. The resulting level of expected utility is clearly seen to be less than that for full insurance

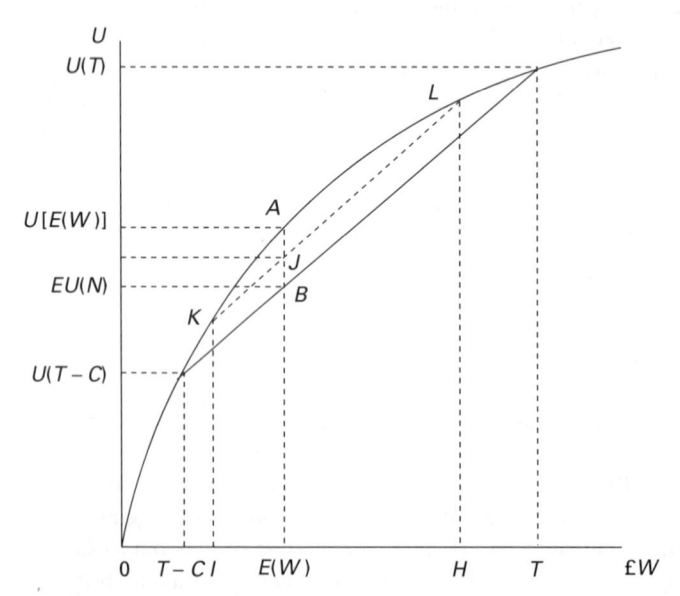

Figure 5.3 The superiority of full insurance

and more than that if uninsured. Thus partial insurance improves the expected utility of the risk averse individual but not as much as is achieved by full insurance; this is not surprising, since partial insurance only partially frees the individual from risk.

The above argument holds for any level of partial insurance. The reader should be able to modify the argument to show that *more-than-full insurance* at fair odds (that is, where the compensation if the car is stolen, Y, exceeds the value of the car) leads to a reduction in expected utility below the level achieved under full insurance. The intuition behind this result is obvious: more-than-full insurance means taking a risk at fair odds (this time with the higher wealth level occurring when the car is stolen) and so decreases expected welfare compared to full insurance for a risk averse individual. The reader should also be able to show that an uninsured, risk averse individual would not wish to engage in a gamble at fair odds which would offer him a payment in the event that his car was *not stolen*; intuitively, such a gamble at fair odds would be risk increasing and reduce his welfare.

The state space representation

Before introducing problems of asymmetric information in the following chapters, it is worthwhile at this stage to introduce the *state–space representation* of the above symmetrical information insurance problem. This may be done by using Figure 5.4.

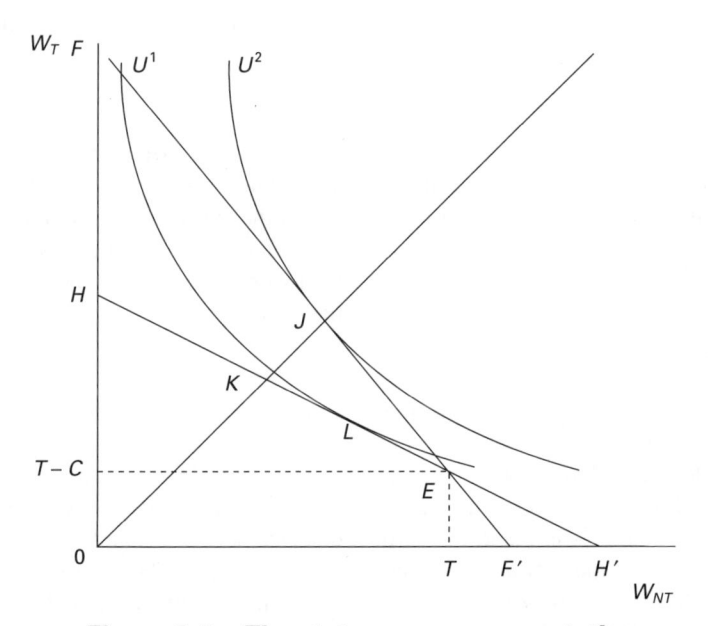

Figure 5.4 The state–space representation

86   The Insurance Market

Figure 5.4 plots the individual's wealth in the case of theft of his car, W_T, along the vertical axis and his wealth in the no-theft case, W_{NT}, along the horizontal axis; hence the state–space jargon, since the axes show wealth in the two possible states of theft or no-theft.

Point E in the figure is the *endowment point*. It represents the state payoffs with which nature endows the individual. Thus at E, W_T is equal to $(T - C)$ and W_{NT} is equal to T. The 45° line through the origin is the *full insurance* or *certainty line*; given equal scales along both axes it represents all points at which the wealth level of the individual is the same in each state of the world.

The line FF' through the point E is the *fair odds line*. The fair odds line is the locus of all points in the state–space that are available to the insuree buying insurance at the fair odds rate. If the individual purchases insurance at fair odds the following holds:

$$pY = X \tag{5.5}$$

where Y is any level of compensation if theft occurs and is no longer necessarily equal to C. Equation (5.5) is just the generalisation of equation (5.4) and represents the competition constraint for levels of compensation different from the full value of the car. For an insured individual, W_T equals $(T - C - X + pY)$ and W_{NT} equals $(T - X)$, so that his expected wealth, $E(I)$, is given by:

$$\begin{aligned}E(I) &= p(T - C - X + Y) + (1 - p)(T - X)\\ &= p(T - C) - p(X - Y) + (1 - p)(T) - (1 - p)(X)\\ &= p(T - C) + (1 - p)(T) - X + pY\end{aligned} \tag{5.6}$$

Substituting for X equals pY from equation (5.5) for insurance at fair odds into equation (5.6), we see that:

$$E(I) = p(T - C) + (1 - p)(T) = T - pC \tag{5.7}$$

Comparing equations (5.7) and (5.2) reveals that the expected wealth under insurance at fair odds is equal to the expected wealth if uninsured, $E(W)$. Hence, purchasing insurance at fair odds does not affect the expected wealth of the individual, as is obvious from noting that, from equation (5.5), under insurance at fair odds the premium, X, equals the expected value of the compensation, pY. It follows that the equation of the fair odds line is given by:

$$pW_T + (1 - p)W_{NT} = E(W) \tag{5.8}$$

Since equation (5.8) is derived using equation (5.5), it is also sometimes known as the zero profit or competition or fair odds constraint.

An alternative derivation of equation (5.8) is as follows. Consider the slope of the fair odds line between points E and F. The horizontal difference between E and the vertical axis is T and the vertical distance between E and F is the compensation minus the premium for a premium of T; that is, $T/p - T$, or $T(1 - p)/p$. Hence the slope, dW_T/dW_{NT}, is $-(1 - p)/p$. This result combined with the knowledge that W_T at F equals $T/p - C$ (which may be calculated by subtracting a premium of T and the value of the car, C, from the initial wealth of T and adding a compensation of T/p) yields, after a little manipulation, equation (5.8).

Points along the fair odds line to the right of E represent points to which the individual could move if he gambled at fair odds; that is, if he made a certain payment in return for a payoff if his car was *not stolen*. Points along the fair odds line to the left of E represent points to which he could move if he purchased insurance at fair odds; full insurance is shown at the intersection of the fair odds line and the full insurance line at J, while points to the right of J and to the left of E represent partial insurance, and points to the left of J represent more-than-full insurance.

Now consider the two indifference curves, U^1 and U^2, shown in Figure 5.4. Of the two, it is clear that U^2 represents the higher level of expected utility to the individual. This is obvious since U^2 cuts the full insurance line, at J, further from the origin than does U^1. The individual would clearly prefer full insurance at the point J to the point where U^1 cuts the full insurance line, since he is guaranteed a higher payoff at point J than at the alternative (unlabelled) full insurance point on U^1.

The indifference curves slope downwards from left to right as shown, since to maintain a constant level of expected utility along an indifference curve it is necessary to increase the level of wealth in the no-theft state to compensate for a reduction in wealth in the theft state, and vice versa. The indifference curve U^2 touches the fair odds line tangentially at the full insurance point J, showing that the slope of the indifference curve equals that of the fair odds line at full insurance. This result is a consequence of the fact that if offered insurance or gambles at fair odds the risk averse individual will choose to fully insure, as we showed above with the help of Figure 5.3. Combined with the downward slope of the indifference curve, this latter result verifies the convex shape of the indifference curves, since utility is higher for a risk averse individual at J than at E.

It is straightforward to verify with calculus that the indifference curve U^2 touches the fair odds line tangentially at J. The equation of the indifference curve is given by:

$$E(U) = pU(W_T) + (1 - p)U(W_{NT}) = U^2 \tag{5.9}$$

Totally differentiating equation (5.9), we find the slope of the indifference curve to be given by:

$$dW_T/dW_{NT} = -(1 - p)U'(W_{NT})/pU'(W_T) \tag{5.10}$$

that is, the *marginal rate of substitution* between wealth in the theft state and wealth in the no-theft state along the indifference curve is just the negative of the probability weighted marginal utility of wealth in the no-theft state over the probability weighted marginal utility of wealth in the theft state.

At point J, or any other point on the full insurance line, the level of wealth, and hence also the marginal utility of wealth, is equal in each state, so that from equation (5.10) we see that the slope of any indifference curve where it cuts the full insurance line is equal to $-(1-p)/p$. The slope of the fair odds line, which represents the *marginal rate of transformation*, has already been shown to be equal to $-(1-p)/p$ (alternatively, total differentiation of equation (5.8) yields the desired result), thus the slopes of the indifference curves at points of full insurance are equal to the slope of the fair odds line. Hence the indifference curve, which is tangential to the fair odds line, touches it at point J and shows that full insurance would be chosen by a risk averse individual offered insurance at fair odds.

Point J will be produced as the market equilibrium under symmetric information in our example. Insurance companies will not be prepared to offer insurance at favourable odds as this would yield them losses. While they would be prepared to offer insurance at unfair odds (such as at the full insurance point K in Figure 5.4), this would involve them making supernormal profits (with X greater than pY) and so competition will force them to offer insurance at fair odds, which represent the best odds they can offer insurees while not making losses. Insurees offered insurance at fair odds will maximize their level of expected utility by choosing full insurance and reaching the highest possible indifference curve obtainable along the fair odds line.

Finally, notice that if an individual were offered insurance at unfair odds $(p' > p)$ along the line HH', he would choose partial insurance or no insurance at all. The case of partial insurance is shown in the figure by the tangency between the indifference curve U^1 and HH' at point L. The slope of the indifference curve at L is equal to the slope of HH' (that is, $-(1-p')/p'$), which is less than the slope of the fair odds line. That the slope of the indifference curve for points of partial insurance is less than the slope of the fair odds line may be seen by examination of equation (5.10) and noting that at points below the full insurance line, such as L, the value of W_{NT} is greater than W_T, so that $U'(W_t)$ is greater than $U'(W_{NT})$, given the diminishing marginal utility of wealth; the slope of the indifference curve is therefore less than $-(1-p)/p$ at such points. Indeed, it would be possible for the marginal rates of transformation and substitution to be equated along an unfair odds line at a point to the right of E; that is, at a point where the odds, although unfair for insurance, represent favourable odds for gambles. However, no company would offer an individual such gambles since to do so would involve the company in making expected losses.[5] In this case, the individual would choose to stay at the endowment point rather than purchase insurance at the unfavourable odds.

Having examined the insurance market under symmetric information, we are now ready to introduce problems of asymmetric information, beginning in the next chapter with the problem of moral hazard with hidden actions. We close the present chapter with some recommended reading and an exercise.

5.4 Recommended reading

The method of calculating expected utility we have presented in this chapter was introduced by von Neumann and Morgenstern (1944). For an excellent and

5.5 Problems

Problem 5.1
Imagine Tracey has total wealth, T, equal to 22. Part of her wealth is her motor car, which she values at C equal to 4. The probability of theft of her motor car is 1/2. Her utility from wealth is given by U equals $\ln(W)$, where W is her wealth. Tracey is offered insurance at fair odds.

(a) What is the equation of the fair odds line?
(b) Show that Tracey is risk averse. Set up and solve the problem of maximizing her expected utility subject to the fair odds line. How much insurance premium will she pay, what compensation will she receive if a theft occurs, and what will be her wealth in the theft and no-theft states of the world?
(c) Illustrate your answer using a state–space diagram.

Problem 5.2
For the above problem, imagine that Tracey is offered insurance at unfair odds by her local insurance broker, Ben. Ben sets a price of p' (greater than 1/2) per unit of compensation.

(a) What is the equation of the unfair odds budget line facing Tracey? Show that anywhere along this line W_T is given by $[T - C + (1 - p')Y]$ and W_{NT} is given by $(T - p'Y)$.
(b) Explain the meanings of the terms $(1 - p')Y$ and $p'Y$ if Tracey takes out insurance.
(c) Tracey complains that the insurance is far too expensive and that at those odds she would rather gamble, making a payment to Ben of B in return for a payment from him of R in the case where her car was not stolen. Show that if B and R are determined by the unfair odds line found in (a) above, the terms B and R satisfy $B = (1 - p')R$.
(d) In the interests of good customer relations Ben agrees to let Tracey take a gamble. Illustrate Tracey's gamble on a state–space diagram.
(e) Find a formula for Ben's expected profits from accepting Tracey's bet. Why would Ben be unwise to accept too many deals like the one he made with Tracey?

CHAPTER 6

Insurance and the Hidden Action Problem

6.1 Overview

We may change the model of the market for insurance discussed in Chapter 5 by assuming that the probability of theft of the individual's car is no longer exogenous but depends upon the individual's actions, say whether he locks his car or not on leaving it. In a full information world the optimum insurance contract would make compensation contingent upon the probability of theft, with a higher price per unit of compensation the higher the probability of theft. In this case, the market for insurance would continue to function efficiently, since if an individual chose to behave in a more risky way, he would pay for this behaviour by paying a higher price per unit of compensation.

We introduce the hidden action problem into this market by assuming that the individual's actions with regard to the care he exercises are unobservable to the insurance company.[1] In this case, it becomes impossible to make the compensation contingent upon the probability of theft, or equivalently upon the amount of care taken by the individual, and this leads to the insurance company responding to the problem by adjusting the terms of the insurance contract it is prepared to offer to its customers. In particular, we shall see that the insurance company is unable to offer the insuree the full insurance that would be the market equilibrium contract in the absence of the hidden action problem. The analysis is presented in section 6.2. Section 6.3 recommends some further reading and section 6.4 presents some problems.

6.2 Insurance and the hidden action problem

Consider the individual wishing to insure his car against theft, as in the previous chapter. Assume now, however, that the individual may choose between two alternative levels of care, *high* and *low*. If he exercises a high level of care the probability of theft of his car is p and if he exercises a low level of care the probability of theft rises to p'. We assume that taking the high level of care imposes a small cost of β upon the individual compared to exercising the low level of care.

We may analyse the insurance market with the hidden action problem using Figures 6.1, 6.2 and 6.3. All three figures are state–space representations of the same problem, but it is clearer to use three diagrams rather than to have too many lines and curves on one diagram.

First consider Figure 6.1, which shows two fair odds lines FF' and HH'; the former is appropriate for the case where the individual chooses to exercise a high level of care and the latter is appropriate when he chooses a low level of care. Applying the analysis from Chapter 5 we know that the slopes of these two lines are $-(1-p)/p$ and $-(1-p')/p'$ respectively.

In Figure 6.1 it is necessary to consider two sets of indifference curves for the same individual, since his level of expected utility at any point in the state–space diagram depends upon whether he exercises the high or the low level of care. Thus any point in the diagram is cut by two indifference curves, one for each level of care. Using equation (5.10) we can see that the relative slopes of the indifference curves depend upon the terms $-(1-p)/p$ and $-(1-p')/p'$, where, since p' exceeds p, the slope of the high-care indifference curve through any point is steeper than that of the low-care indifference curve through that point. Thus, considering the two indifference curves U^1 and U^2, which pass through point J at the intersection of the full insurance line and the high-care fair odds

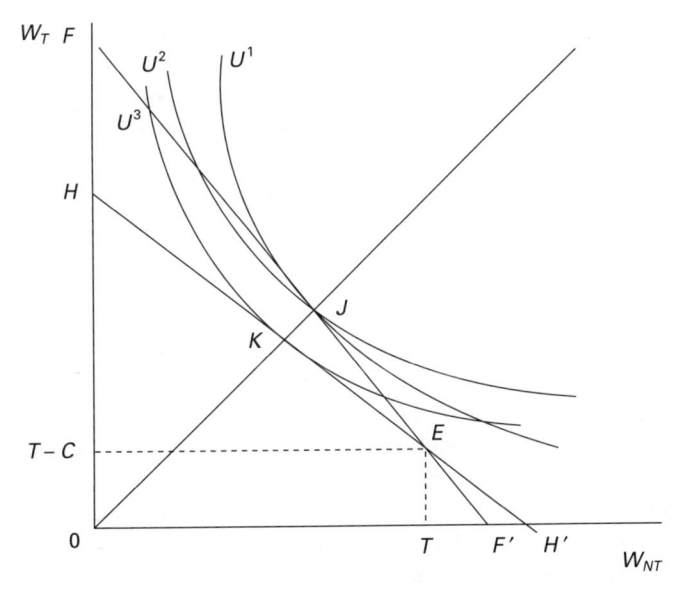

Figure 6.1   No full insurance at fair odds

line, U^1 is relevant if the individual exercises a high level of care and U^2 if he exercises a low level of care.

The intuition for the relative slopes of these two indifference curves is straightforward. Consider a movement down the indifference curves away from point J. Such a movement involves accepting less wealth if theft occurs and gaining a higher level of wealth if the no-theft state occurs. If the individual is exercising a high level of care then the chance of theft is lower than if he is exercising a low level of care, hence he will be prepared to give up a higher amount of wealth in the theft state for a similar gain in the no-theft state if he exercises the high rather than the low level of care, because then the chance of actually making that loss is lower and the chance of making the gain is higher. In other words, a person in a very risky situation is willing to pay more for a unit of compensation than a person in a safer situation. Thus the indifference curve through point J for the high level of care is steeper than the indifference curve through point J for the low level of care. A similar logic applies for any point in the state space; when indifference curves intersect the steeper of the two is for the high level of care.

Since the indifference curves U^1 and U^2 both pass through the full insurance point J we know that the level of utility U^2 is equal to U^1 plus β. This follows since at J the individual receives the same certain level of wealth regardless of the level of care he exercises and therefore receives the same utility from wealth regardless of the level of care chosen. However, he gains β more utility from choosing the low rather than the high level of care.

We may deduce from the fact that U^2 exceeds U^1 that the insurance contract will no longer allow the insuree to choose full insurance at fair odds for a high level of care. This follows, since if the individual were able to so choose, he would locate at point J in the diagram and would then choose the level of care that maximizes his utility; that is, the low level of care. However, if he exercises the low level of care at point J, he produces expected losses for the insurance company, since the line FF' is the fair odds line for the high level of care.[2]

The individual could, of course, be offered insurance along HH' at fair odds appropriate to him choosing the low level of care. We examine the possibilities in this case using Figure 6.2.

If he chose the low level of care, he would choose to locate at point K in Figure 6.2; that is, where the low-care indifference curve U^3 is tangential to HH'. However, if offered insurance anywhere along HH', he would choose to exercise the high level of care and take only partial insurance at the point Q, where the high-care indifference curve U^4 is tangential to HH'. He chooses partial, rather than full, insurance, since when he exercises the high level of care the odds are unfair along HH'. In terms of the diagram it is possible to see that U^4 represents a higher level of utility than U^3, since it cuts the full insurance line at R to the north-east of K. At R, the individual is better off, even if he exercises the high level of care, than at K, since he has a higher fully insured level of wealth, which gives him more utility than he would get by moving to K (and exercising the low

The Hidden Action Problem 93

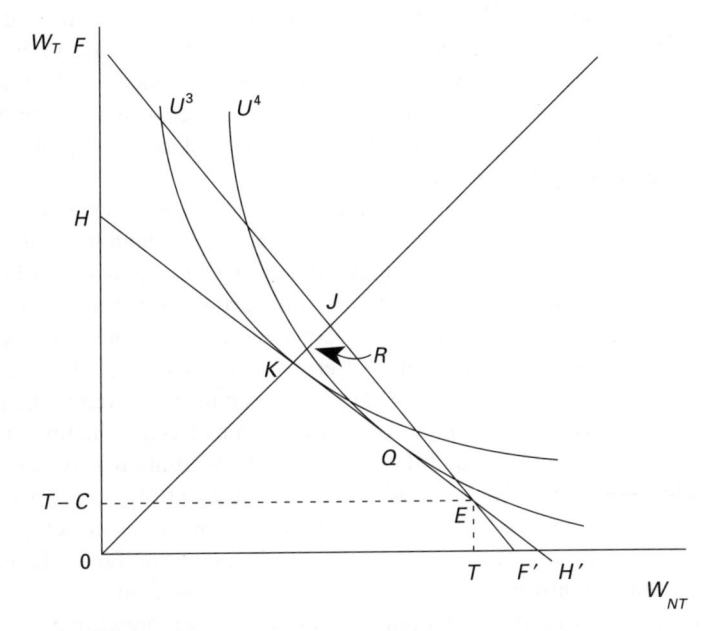

Figure 6.2 Partial insurance at unfair odds

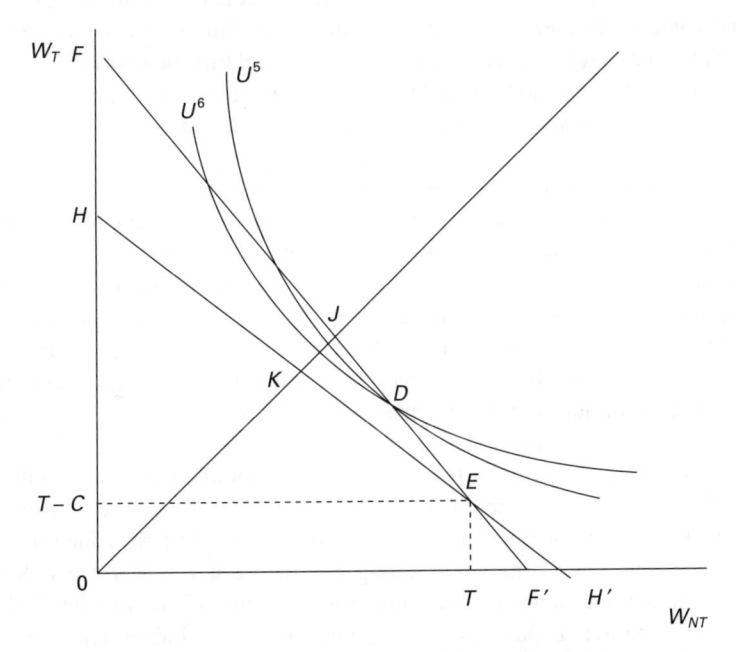

Figure 6.3 Hidden action and partial insurance

level of care to gain only the small amount of utility of β). Since the individual is as well off at Q as he is at R when exercising the high level of care, it follows that U^4 exceeds U^3. Alternatively, a comparison of the choice of a high level of care at Q or a low level of care at K indicates that in choosing the former, the insuree makes the small loss of β but he gains by reducing the probability of theft and purchasing less insurance at Q than at K.

The equilibrium in a competitive insurance market, however, will not be at point Q. This follows, since at Q the insuree is taking the high level of care but being charged a price per unit of compensation of p' in excess of the fair odds price of p; thus an insurance company offering partial insurance at Q would be making supernormal profits. These supernormal profits would be competed away by other insurance companies offering better insurance deals until the supernormal profits were eliminated. The insurance deals offered to the insuree would be improved relative to the contract at Q in two ways: the price per unit of compensation would fall and the amount of compensation purchased at the lower price would rise. This process would continue until the price per unit of compensation was reduced to the rate p appropriate for the insuree choosing the high level of care; that is, the equilibrium will be along the fair odds line FF'. We know already, however, that the full insurance point J on FF' cannot be an equilibrium. The equilibrium will, instead, be at a point representing partial, but almost full, insurance, such as at point D in Figure 6.3.

Figure 6.3 again shows the two fair odds lines FF' and HH'. Any point along FF' where the individual chooses the low level of care may be said to represent an insurance contract which does not satisfy incentive compatibility. This follows since the line FF' is the fair odds line, given that the individual exercises the high level of care; when he chooses the low level this means that the contract gives him an incentive to behave in a way incompatible with the way in which the insurer would want him to behave under that contract if it is not to make losses.

We know that at point J the individual would choose to exercise the low level of care as he gains β compared with choosing the high level of care. Points along FF' to the north-west of J only provide more inducement for the individual to choose the low level of care, since at such points he is better off if his car is stolen, since W_T exceeds W_{NT}; this explains why insurance companies usually take care to avoid allowing an insuree to become more than fully insured, since once in this situation the insuree becomes tempted to bring about the hazard against which he has purchased insurance.[3] Such points therefore represent incentive incompatible contracts.

At E, however, we assume that the individual would choose to take the high level of care, since we assume that β is small and that if he were uninsured the individual would be prepared to pay the cost of β in order to reduce the probability of theft. Thus somewhere along FF' between E and J the individual would be indifferent between choosing either the high or the low level of care. This point is shown as point D on the figure, where the indifference curves U^5 for the high level of care and U^6 for the low level of care intersect. The individual is indifferent at this point between choosing the high or the low level of care, so that U^5 equals U^6. Since U^5 is for the high level of care, it follows that it must cut the full insurance line to the north-east of the point where U^6 cuts it, so that the extra fully insured level of wealth compensates for the loss of β involved in choosing the high level of care.

If we make the assumption of epsilon altruism (that is, when indifferent between two actions, an agent chooses that action which is better for the principal), then it follows that points to the north-west of D along FF' represent incentive incompatible contracts while point D and points to the south-east of D along FF' represent incentive compatible contracts. Thus at D, W_T is less than W_{NT} by just enough to make the individual wish to avoid the theft occurring to make it worth his while to pay the cost of β associated with choosing the high level of care. For points along FF' to the south-east of D, the gain in expected utility from taking the high rather than the low level of care exceeds β, and for points to the north-east of D the gain in expected utility is less than β.

Since we assume β to be a small number, it follows that point D will be near point J and represents almost-full insurance. Point D represents the most insurance, in the sense of the largest compensation, at fair odds that the insurance company can offer to the individual while maintaining incentive compatibility. The individual will prefer point D to any other point along FF' at which he would choose to be careful, and so competition will drive insurance companies to offer him the insurance contract defined by point D.

In terms of real world contracts, the insurance contract at D may be viewed as one which offers only partial compensation against the insured hazard, or as one in which the insuree agrees to pay an *excess* or *deductible*; that is, where he agrees to pay for, say, the first £100 of any loss. Such contracts may be a response to the problems of asymmetric information we have analyzed here, or those that we will analyse in the next chapter, or, since handling insurance claims is costly, they may be a device to economise on the number of claims processed by inducing customers not to make small claims.

The exact amount of insurance provided at point D may be evaluated by solving the maximisation problem of the individual subject to the constraints that insurance must be offered along the fair odds line and that the contract must be incentive compatible. The incentive compatibility constraint is expressed as:

$$pU(W_T) + (1-p)U(W_{NT}) - \beta \geq p'U(W_T) + (1-p')U(W_{NT}) \tag{6.1}$$

where $W_T = (T - C - pY + Y)$ and $W_{NT} = (T - pY)$. The left-hand side of constraint (6.1) represents the expected utility if the high level of care is taken, and the right-hand side is for the low level of care. Given that the individual chooses the high level of care, the objective to be maximised is clearly the left-hand side of Equation (6.1).

Since the solution at point D occurs on the fair odds line where the expected utility level of the insuree if he takes the high level of care is equal to the expected utility level if he takes the low level of care, it is possible to determine the optimum contract by treating constraint (6.1) as an equality and solving simultaneously with the equation of the fair odds line to determine W_T and W_{NT} and hence to determine the terms of the contract at D.

Notice that in a welfare sense the equilibrium at point D is inferior to the individual taking the high level of care under full insurance at point J. The insurance company breaks even in either case but the insuree taking the *high* level of care would prefer full insurance at J to almost-full insurance at D. However, if offered full insurance at J, he would choose the low level of care and impose expected losses on the insurer. Hence the very fact that the insuree is able to choose the low level of care and unable to commit himself to taking the high level of care, or, equivalently, is unable to take a contract which is contingent upon his hidden actions, works to his disadvantage. Since β is small and D near to J, the welfare loss is small, but it does exist. Thus we see that the problem of moral hazard with hidden actions leads to an equilibrium at D which is inferior to that which could be achieved at J if information on actions was freely available. The logic is straightforward: the insurance company cannot offer full insurance at J and break even, so it offers the partial insurance contract at D, under which the insuree is exposed to just enough risk to make it worth his while to exercise the high level of care.

Finally, notice that although the insurer does not wish to see the insuree take out full insurance along FF', the insuree would wish to be fully insured at J. Why, then, does the insuree not take two contracts with different insurance companies and insure half the value of his car with each company? The answer is that if he did so and became fully insured he would no longer have the incentive to take the high level of care, so insurance companies prevent this sort of behaviour by making it a condition of insurance contracts that the risk be insured by only one company or that any other insurance against the same risk be declared to them by the insuree.

Having seen how the problem of moral hazard with hidden actions affects the insurance market we turn in the next chapter to see how the selection problem may have rather more serious effects upon the market. The remaining sections of this chapter present some recommended reading and a problem.

6.3 Recommended reading

The key article on the problem of moral hazard with hidden action and the implications for contract design is Grossman and Hart (1983).

6.4 Problems

Problem 6.1
Reconsider problem 5.1 on page 89, now assuming that the probability of Tracey's car being stolen is 1/2 if she exercises a high level of care and 3/4 if she exercises a low level of care. Other things being equal, Tracey prefers to

exercise the low level of care, since she incurs disutility of β if she exercises the high level of care.

(a) Explain why no full insurance contract will be offered in market equilibrium if the level of care exercised by Tracey is non-contractible.
(b) Set out, but do not solve, the maximisation problem which could be used to determine the equilibrium solution if Tracey purchases insurance in a competitive market. Be careful to specify the incentive compatibility and competition constraints.
(c) Assuming the problem in (b) above has the usual solution, it is possible to determine the equilibrium outcome simply by solving simultaneously two equations. State which two equations these are and find the solution. Explain your answer with the aid of a diagram.

CHAPTER 7

Insurance and the Selection Problem

7.1 Overview

We may introduce the selection problem into the insurance market by imagining that there are two types of risk averse customer who wish to insure their cars against theft. The two types of customer are identical in every way but one: the *safe* type has an exogenously given probability of theft of p, while the *risky* type has an exogenously given probability of theft of p', which is greater than p. The insurees know their own type, but insurance companies are unable to distinguish between the two types and have to take this asymmetry of information into account when determining the contracts they offer to potential customers.[1] We shall see that the resulting market equilibrium reduces the welfare of the safe types below that which they could achieve in a full information world.

Section 7.2 sets out the analysis under full information to serve as a benchmark for the analysis of asymmetric information in the following sections. Section 7.3 deals with the case where insurers offer a common contract to all insurees. This is known as the *pooling case* and the common contract is known as the *pooling contract*, since all customers are pooled together. It is shown that a pooling contract imposes costs on the safe customers and provides benefits to the risky ones compared to the full information equilibrium and that, in the extreme, no pooling contract is feasible (in the sense that no company can offer a pooling contract and avoid losses) and adverse selection occurs with only the risky customers being offered insurance. Whether or not a pooling contract is feasible, it is shown that it is impossible for a pooling contract to produce a Nash equilibrium. Section 7.4 therefore looks at the possibility of a *separating equilibrium*, where the insurers offer two contracts and the different risk categories separate themselves by their contract choice; that is, they each accept a different contract. It is shown that a separating contract may in some circumstances lead to a Nash equilibrium, but in other circumstances may not. In the latter case it becomes necessary to consider other concepts of equilibrium and

7.2 Insurance and different risk categories under full information

The case where insurers are able to distinguish between customers from each of the two risk categories is straightforward. Assuming, as in the previous chapters, that insurance companies are risk-neutral profit maximisers operating in a perfectly competitive market, it is easy to see that the market equilibrium will be characterised by each type of customer choosing full insurance at the fair odds rate appropriate to his risk category. This equilibrium is illustrated in Figure 7.1.

The figure shows two different fair odds lines, FF' for the safe customers and HH' for the risky customers, where the endowment point E is common to both types. The line HH' represents the more expensive premium per pound of compensation necessary for insurers to break even on contracts with the risky type of customer compared to the safe type, as is easily seen by comparing the full insurance point on HH' with its counterpart on FF' which shows a higher level of fully insured wealth and indicates a lower cost of purchase of insurance cover. The slopes of the fair odds lines may, of course, be evaluated in the same way as discussed in Chapter 5 above.

Both types of customer have the same preferences but their indifference curves have different slopes at any point in the state–space diagram since they face different probabilities of having their cars stolen. The risky type are more likely to have their car stolen and, therefore, from any point in Figure 7.1 they would require a smaller increase in wealth in the theft state in compensation for

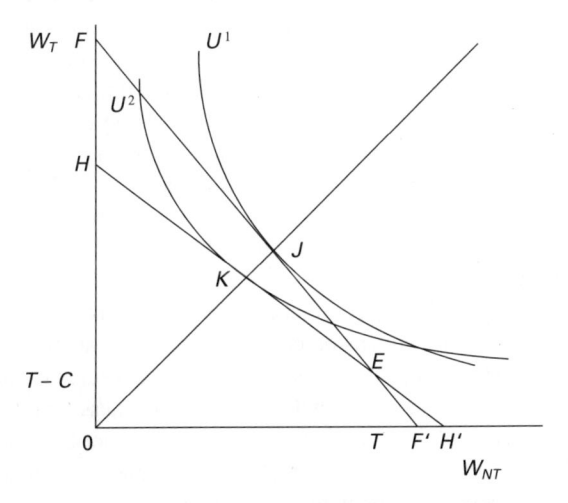

Figure 7.1 Insurance and different risk groups

a given decrease in wealth in the no-theft state than would their safe counterparts; hence, at any point where the indifference curves for the two types of customer cross, the indifference curve for the safe customer is steeper than that for the risky customer, as illustrated by the indifference curve U^1 for the safe customer being steeper than the curve U^2 for the risky customers at the point where they intersect. More formally, the slopes of the indifference curves for the two types of customer at any point may be compared by using equation (5.10) on page 87. It may also be shown that our assumptions imply that any given indifference curve for a safe customer cuts any given indifference curve for a risky customer once at most; this property is known as *the single-crossing property*.

In equilibrium, each type of customer will be offered insurance along the appropriate fair odds line. No insurer would wish to offer the risky customer insurance along the safe fair odds line since this would lead to losses, while competition for custom ensures that each type is offered insurance at the appropriate fair odds. Thus risky customers are offered insurance at a premium of p' and choose full insurance at point K on HH', and their safe counterparts are offered insurance at a premium of p and choose full insurance at point J on FF'. Each type of customer maximises his utility, given that the insurance is offered at fair odds and the analysis is really a simple extension of that of section 5.3.

Given this full information benchmark, we turn to the asymmetric information case, dealing first of all with the pooling case.

7.3 Pooling together different risk categories

We now introduce the asymmetry of information and assume that insurers are unable to distinguish between safe and risky customers. In this case, the contracts offering location at points J and K in Figure 7.1 will no longer be offered together, since both safe and risky types would prefer to accept the cheaper insurance and locate at point J; that is, the risky customers would accept the contract designed for the safe type and impose losses on the insurers by doing so. The insurers must therefore respond to the asymmetry of information and take it into account in determining which contracts to offer. In this section we consider the case where the insurers must offer only one type of contract to any customer; that is, they must pool the customers. This assumption will be relaxed in the next section.

If insurers must offer only one contract it follows that this contract cannot be at fair odds appropriate to the safe customers. This is so since such a contract would be attractive to both safe and risky customers so that the insurer would break even on business with safe customers but make losses on business with the risky ones, and so make losses overall. The contract will therefore either be at odds appropriate to the risky customers, in the case of adverse selection, or at the *market average fair odds*. Let us deal first with the latter case, using Figures 7.2 and 7.3.

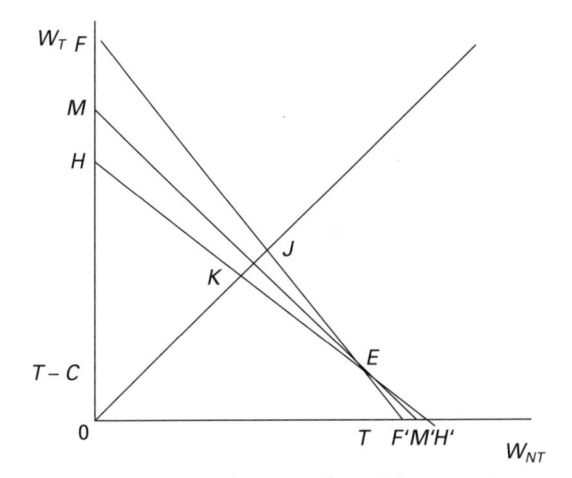

Figure 7.2 The market average fair odds line

Figure 7.2 shows the market average fair odds line MM' cutting through the endowment point E in between the fair odds lines FF' and HH'. The market average fair odds are simply the odds that an insurer could offer to the average customer while breaking even on average as long as the contract was accepted by a random sample of both types of customer. In other words, the premium per unit of compensation, or the *market average fair premium*, is p^M equal to $(n_1 p + n_2 p')/(n_1 + n_2)$, where n_1 and n_2 are the numbers of safe and risky customers, respectively, in the population.

The market average fair premium, p^M, represents unfavourable odds to the safe customers and favourable odds to the risky ones. If able to choose the level of insurance coverage at those odds, the safe customers would choose partial insurance and the risky customers would like to choose more-than-full insurance.[2] The insurer offering contracts along MM' will, however, be driven by competition to offer the pooling contract along MM' which optimises the welfare of the safe customer. This is illustrated at point L in Figure 7.3. The utility of the safe type of customer is maximised at L, as shown by the steeper indifference curve U^1 for the *safe* type touching the line MM' tangentially at L.

Any pooling contract offering location at a point below MM' would offer supernormal profits to the insurer if it attracted both types of customer. Such a contract could not therefore be consistent with equilibrium, since competition would drive the contract terms to be improved until the market average fair odds were being offered.

Any contract to the right of L along MM' could be improved upon by another insurer offering a contract allowing location at L; since both risky and safe customers would prefer the contract at L, and insurers offering such a contract would attract all customers away from insurers offering contracts to

102 The Insurance Market

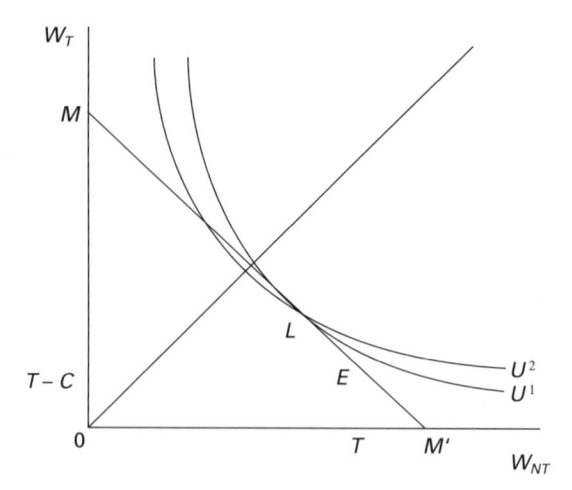

Figure 7.3 The pooling contract

the right of *L*. Any contract along *MM'* to the left of *L* could be improved upon by an insurer offering a contract allowing location at *L*; because only the safe customers would accept the contract at *L* and that contract would produce supernormal profits since it lies below the safe customer's fair odds line. Conversely, the contract to the left of *L* would become loss-making as it would continue to attract only the risky customers and it lies above their fair odds line.

Thus, insurers offering pooling contracts along *MM'* will be driven by competition to offer the contract allowing location at *L* where the utility of the safe customers is maximised. Both types of customer will accept this contract and the insurers will make normal profits. Risky customers would prefer to buy more insurance than the contract at *L* allows, but they would not offer to do so because if they did it would signal their type to the insurer.

No pooling contract like that illustrated in Figure 7.3 may be feasible. In this case, adverse selection occurs and only the risky customers purchase insurance. This case is illustrated in Figure 7.4.

Figure 7.4 shows an indifference curve U^S for the safe type of customer, which is steeper than the market average fair odds line, *MM'*, through the endowment point *E*. Thus, if offered insurance at the market average premium, p^M, the safe type of customer would choose not to insure and to remain at the endowment point in the state space, since expected utility is higher at *E* than at any point to the north-west of *E* along *MM'*. Indeed, it can be seen that the safe customers would prefer to gamble rather than to insure along *MM'*, since the gambles would be at favourable odds; however, such gambles would involve expected losses for the 'insurers' and would not be offered.[3]

The risky customers alone would be prepared to accept contracts offering location along *MM'* to the north-west of *E*, but they will not be offered such

The Selection Problem 103

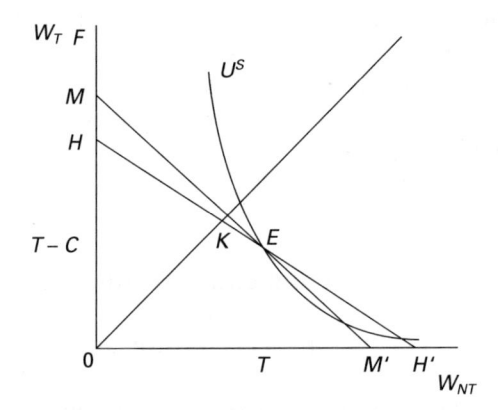

Figure 7.4 Adverse selection

contracts since these would involve expected losses for the insurers, who instead will offer insurance at the premium p' appropriate for risky customers, who will then optimise by choosing full insurance at point K on HH'.

For the adverse selection case to occur, the slope of the safe customer's indifference curve through the endowment point must be steeper than the slope of the market average fair odds line. For this to be the case there must be either a preponderance of risky customers or a big difference in risk associated with the two types of customer, which will make the market average fair odds line relatively nearer to HH' than to FF', or else the customers must not be very risk averse, which will tend to make the indifference curves steep through the endowment point.

The analysis of the pooling contract so far exactly mirrors the analysis of the selection problem in the credit market, with an infinitely elastic supply of credit in section 2.2. In the credit market, we saw that the selection problem either leads to the better customers (from the banks' point of view) paying an average interest rate higher than they would have to pay in a full information world, or being persuaded to leave the market for credit altogether in the case where no feasible interest rate could be charged to allow the banks to break even on loans to a pooled sample of borrowers – in which case, only the riskier projects were funded. The analogous results here are either that the safe customers pay an average premium rate that is higher than they would pay in a full information world (and less than the risky customers would pay under full information) or else they are excluded from the market and only the risky customers purchase insurance at an appropriate rate. In either case, the safe customers are penalised as a result of the asymmetry of information and the presence of risky customers in the population who are unwilling to disclose to the insurers that they are risky since so doing prevents them from purchasing cheaper insurance.

Just as in the credit market case, however, it is possible to argue that the market will respond to the asymmetry of information to move away from the pooling result. In the credit market example we found an alternative form of contract – equity – which solved the selection problem and produced the first best outcome. No such perfect response is possible in the insurance market, although the insurers may, by offering an appropriately designed menu of contracts, be able to offer the safe customers a better deal than that implied by the pooling contract. We analyse this possibility in the next section.

7.4 Separating contracts and equilibrium concepts

No pooling contract is ever a Nash equilibrium contract

As a first step to seeing how and why the insurance market may move beyond offering a pooling contract or producing adverse selection, notice that neither a pooling contract nor the adverse selection result can ever be consistent with a Nash equilibrium. We illustrate why this is so for the case of a pooling contract in the text and leave the interested reader to deal with the adverse selection case.

It is easy to see, with the aid of Figure 7.5, that no Nash equilibrium is possible under pooling. Consider any point, such as N, produced by a pooling contract anywhere along the market average fair odds line, MM'. The outcome at point N is not consistent with a Nash equilibrium, since the indifference curve for the safe customer, U^1, through any point is steeper than that for the risky customer, U^2, through that point. Thus it is always possible for an insurance company to improve upon a pooling contract such as that underlying point N by offering another contract which would allow the customer to locate at a point

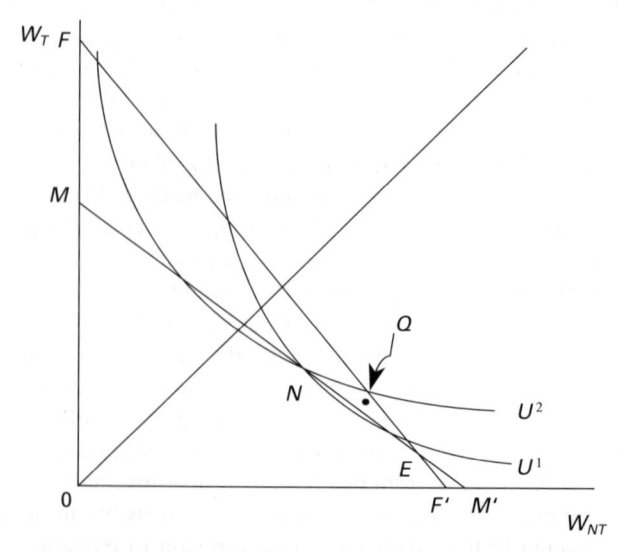

Figure 7.5 **No Nash equilibrium under pooling**

such as Q in the figure. Q lies in the area below the safe fair odds line, FF', and the risky indifference curve, U^2, through N but above the safe indifference curve, U^1, through N. The contract allowing location at point Q would attract safe customers, who prefer Q to N, but would not attract risky customers, who prefer N to Q. Since Q is below the safe fair odds line the contract would produce supernormal profits for the insurance company. Exactly the same logic can be applied to any pooling contract: no such contract is consistent with Nash equilibrium, since the relative slopes of the indifference curves for the two types of customer and the single-crossing property imply that it is always possible to find another contract that will only be attractive to the safe customers and produce supernormal profits.

The act of offering a contract attractive only to customers on whom profits can be made is known, for obvious reasons, as *cream skimming* or *cherry picking*. The cream-skimming contract, however, does not represent a Nash equilibrium any more than the pooling contract from which it skims the cream. This is easily seen to be the case by considering that once the cream has been skimmed from the original pooling contract it becomes loss-making since it only attracts the risky customers at odds that are favourable to them. Therefore the pooling contract will be withdrawn, leaving only the cream-skimming contract. All customers are then attracted to the cream-skimming contract, which either begins to make losses if it is allowing location above the market average fair odds line as shown in Figure 7.5, or, if it allows location below the market fair odds line, will continue to offer supernormal profits even for a random sample of both types of customer. Neither profits nor losses can continue in equilibrium, and competitive forces in either case will cause insurers to offer new contracts different from the cream-skimming contract.

Separating contracts

Although neither the pooling contract nor the cream-skimming contract is consistent with Nash equilibrium it is possible that a pair of *separating contracts* can produce a Nash equilibrium. The separating contracts are a pair of contracts that cause the two types of customer to *separate* or *self-select* or *sort* themselves by choosing between the contracts, so that one type chooses one contract and the other type chooses the other contract. A possible separating equilibrium is illustrated in Figure 7.6.

Figure 7.6 illustrates the outcome of a pair of separating contracts. One contract offers full insurance at the fair odds premium per unit of compensation of p' for the risky customers, so that choosing this contract locates the customer at point K in the state space. The other contract offers only partial insurance, but at the lower fair odds premium per unit of compensation of p which is appropriate for the safe customers and locates those who choose it at point R in the state space. Since the point R lies at the intersection of the safe fair odds line and the unsafe indifference curve, U^2, through K it follows that the risky customers are

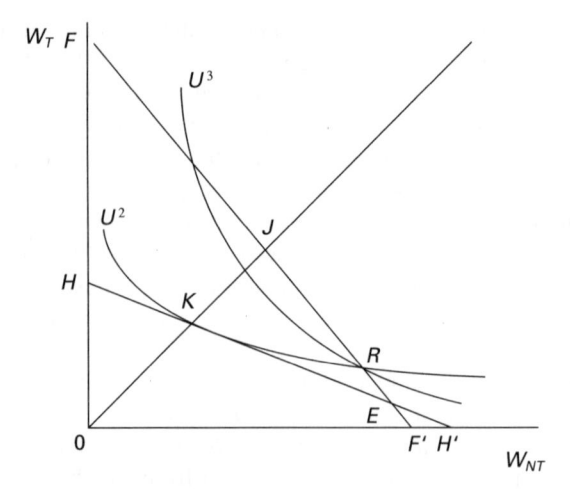

Figure 7.6 A pair of separating contracts

selfishly indifferent between the two contracts since both offer them the same level of expected utility. Making the assumption of epsilon altruism implies that the risky customers will, in fact, choose the full insurance contract in this case, since it allows the insurers to break even, and they prefer that choice to the alternative, which would yield expected losses for the insurers. The safe customers, on the other hand, prefer the point R to the point K (as indicated by R lying on the indifference curve U^3 while K lies below it) and so choose the partial insurance contract which, since it lies on the safe fair odds line, also allows the insurers to break even. The pair of contracts, therefore, induces separating choices and allows the insurers to break even.

Notice that the level of partial insurance offered at the safe fair odds premium is determined by the intersection of the safe fair odds line and the risky indifference curve through K. Any higher level of cover would produce a point to the north-west of R along FF' and would attract the risky customers, while any lower level of cover would mean that it would be possible to offer a slightly higher coverage at a slightly higher premium and attract only the safe customers and make profits. Competition for safe customers therefore drives the level of cover up to that producing point R. No higher level of cover is possible, since it would attract the risky customers and cause expected losses for the insurer along the safe fair odds line.

Formally, the contract underlying point R may be found by maximising the expected utility of the safe type of customer, subject to two constraints. The first constraint is provided by the fair odds line for the safe type of customer, and the second is provided by the *self-selection constraint*. The self-selection constraint states that the risky type of customer must be at least as well off choosing the full insurance contract offered at the risky level of fair odds as choosing the partial insurance contract offered at the lower premium appropriate for safe customers.[4] Thus the self-selection constraint may be written as:

$$U^2 = U(T - p'C) \geq p'U(T - C - pY + Y) + (1 - p')U(T - pY) \qquad (7.1)$$

where U^2 is the level of utility achieved by risky customers who are fully insured at K in the diagram, and the terms after the inequality sign in constraint (7.1) represent the expected utility available to risky customers purchasing insurance at the cheaper premium appropriate for safe customers. Since, as we have shown above, the welfare of the safe customer is maximised at R, where the indifference curve U^2 cuts the safe fair odds line, it is possible to evaluate the contract offered to the safe customers at R by simultaneously solving the equations of the safe fair odds line and the version of constraint (7.1) with the inequality replaced by an equality sign. The assumption of epsilon altruism implies that the resulting level of insurance cover offered to the safe customers is the most that could be offered to them at the safe level of fair odds without inducing the risky customers away from the full insurance contract at K.

Under the separating contracts, the risky customers are maximising their expected utility at K subject to paying the fair odds premium for insurance cover. It follows, therefore, that any other additional contract offering supernormal profits for the insurers must either attract only the safe customers at a premium above the appropriate fair odds premium, or attract both types of customer at a premium above the market average fair odds premium. If such a contract can be found, the separating contracts will not produce a Nash equilibrium, since an insurer would have an incentive to change behaviour and introduce such a contract.

Whether the pair of separating contracts produces a Nash equilibrium or not depends on the position of the market average fair odds line relative to the indifference curve for the safe customer given the separating contracts; that is, the indifference curve U^3 through the point R in Figure 7.6. There are two possibilities, which are illustrated in Figure 7.7.

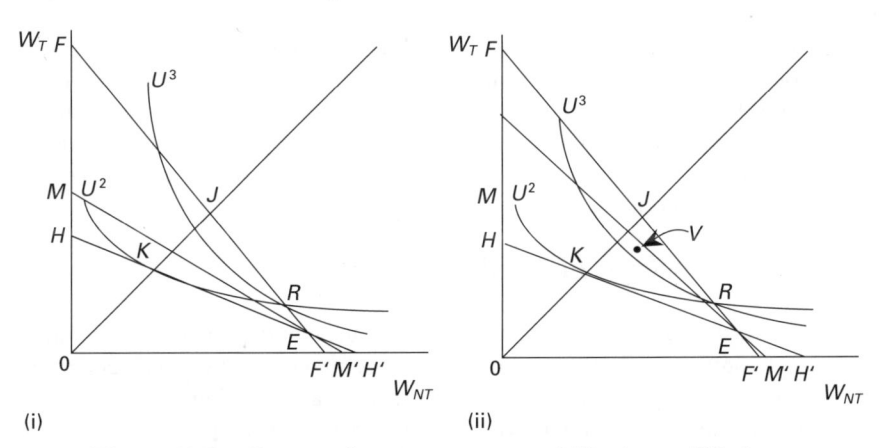

Figure 7.7 Separating contracts and Nash equilibrium

Figure 7.7 (i) shows the case where the market average fair odds line, MM', lies everywhere below the safe customer's indifference curve through R. In this case, any contract capable of attracting the safe customers away from point R would also attract risky customers away from K and lie above the market average fair odds line, thus indicating a premium below the market average fair odds premium and producing expected losses for the insurer. An insurer faced with competitors offering the separating contracts could do no better than to offer those contracts and can find no other contract to offer which promises supernormal profits; the separating contracts, therefore, produce a Nash equilibrium.

Figure 7.7 (ii) differs from Figure 7.7 (i) in the crucial respect that the market fair odds line, MM', is now drawn so that it cuts the safe indifference curve through R. This difference could be due to a higher proportion of safe customers causing the market average fair odds line to rotate about the endowment point E, which is how we have drawn it. Or it could occur if the attitude to risk on the part of the customers was different, causing the indifference curve through R to swivel down and cut the market average fair odds line, as would happen as the customers became more risk averse. If the indifference curve and the market average fair odds line cut in this way, it is possible to find some additional contract to offer that is capable of tempting both types of customer away from the separating contracts and yielding positive expected profits to the insurer. Such a contract is illustrated in Figure 7.7 (ii) of the figure as the contract which allows customers to locate at point V in the state space.

Since V lies above the indifference curves U^2 and U^3 the contract attracts both types of customer away from the separating contracts. Further, since V lies below MM' the contract charges a premium higher than the market average fair odds premium, thus yielding positive expected profits to the insurer. An insurer faced with competitors offering the separating contracts will not maximise profits, given the actions of his competitors, by offering the separating contracts, but will do better to offer the contract allowing customers to locate at point V; the separating contracts, therefore, do not produce a Nash equilibrium in this case.

The contract allowing location at point V is a pooling contract. We saw in the previous section that no such contract ever produces a Nash equilibrium in this market. It follows, then, that no Nash equilibrium exists in the case illustrated in Figure 7.7 (ii); neither the separating contracts nor any pooling contract produce a Nash equilibrium. It becomes necessary, therefore, to consider alternative concepts of equilibrium.

Alternative concepts of equilibrium

When a Nash equilibrium does not exist it means that there is never a contract, or set of contracts, which represents the best contract, or set of contracts, for an insurance company faced with other insurers already offering that contract, or set of contracts. It is natural, however, for the insurance company contemplating introducing a new contract, or set of contracts, to consider the reactions of its rivals to its actions. Let us call the insurance company contemplating this action the *defector* and the contract, or set of contracts, it is contemplating the *defection*.

Considering reactions in this way has produced two important alternative concepts of equilibrium, the *Wilson* equilibrium and the *Reactive* equilibrium. Although both concepts are similar in considering rivals' responses, they lead to quite different conclusions.

Consider first the Wilson equilibrium. The Wilson equilibrium is based on the idea that a defector will only introduce the new contract, or set of contracts, if it considers that the defection will not become unprofitable once initial contracts that are made unprofitable as a result of the defection are withdrawn. This idea leads to the pooling contract producing the Wilson equilibrium.

The pooling contract underlying point V in Figure 7.7 (ii) is clearly not an equilibrium since it yields supernormal profits. Competition would therefore drive insurers to offer a pooling contract on the market average fair odds line at the point where it is touched tangentially by an indifference curve of the safe type of customer. This contract was discussed in Section 7.3 above and illustrated in Figure 7.3, where point L is produced by the pooling contract which maximises the utility of the safe type of customer along MM'. The defection from this contract is the cream-skimming contract which attracts only the safe customers.

Since the contract at L was maximising the utility of the safe customers along the market average fair odds line, the cream-skimming defection must offer them a point above the market average fair odds line; that is, it must charge a premium per unit of compensation less than the market average fair odds premium. However, once the defection is introduced, the original pooling contract becomes loss making, since it only attracts the risky type of customer. The original contract will then be withdrawn and the defection will attract both types of customer and become loss-making since it charges a premium below the market average fair odds rate. The logic underlying the Wilson equilibrium then implies that the defection would not be introduced in the first place, since any potential defector from the original situation with the pooling contract would realise that the defection would become unprofitable once contracts it turned into loss-making contracts were withdrawn. The Wilson equilibrium is therefore the pooling equilibrium in the case where there is no Nash equilibrium. The reader ought to consider the case where the separating contracts produce a Nash equilibrium and should be able to see that in this case there is no defection that would offer profits even if the original contracts were not withdrawn: the separating contracts are in this case, then, consistent with the Wilson as well as the Nash equilibrium.

Now consider the Reactive equilibrium. This is based upon the idea that a defection will only be introduced if it does not become loss-making when another contract, or set of contracts, known as the *reaction*, is introduced. In this case, the separating contracts allowing location at points K and R in Figure

7.7 (ii) produce the Reactive equilibrium. This is so since the defection underlying point V will be made loss-making by competition driving the premium down until the market average fair odds line is reached, and then the cream-skimming contract will be introduced as the reaction to the pooling contract. Hence the defection will not be introduced in the first place and the separating contracts produce the Reactive equilibrium. As for the Wilson equilibrium, the reader ought to be able to see that when a Nash equilibrium exists it is also the Reactive equilibrium.

The Wilson and Reactive equilibria are both based on the idea of considering rivals' responses.[5] Unfortunately, they produce very different equilibria; when no Nash equilibrium exists, the Wilson equilibrium is a pooling equilibrium and the Reactive equilibrium is a separating equilibrium. This seems to be rather unsatisfactory since it leaves us having to choose between two quite different possibilities. The difference results because the Wilson equilibrium is based on rivals *withdrawing* contracts in response to a defection and the Reactive equilibrium is based on rivals *introducing* contracts in response to a defection. Which of the two concepts is likely to be more relevant in the real world may therefore depend upon how quickly rivals may either introduce or withdraw contracts. If it is possible to withdraw contracts very quickly while introducing them takes a long time, then perhaps the Wilson equilibrium is the better concept, or vice versa. For the moment, however, it seems that further work is needed to consider the appropriate equilibrium concept when there is no Nash equilibrium.

Finally, notice that under the separating equilibrium, whether Nash or Reactive, the presence of the risky customers exerts a negative externality on the safe customers who are offered only partial insurance rather than the full insurance they could obtain in a full information world or in a world where the risky customers did not exist. Since the risky customers do not gain anything, but achieve the same level of insurance at the same price as in a full information world, this externality is said to be *dissipative*; that is, it is only harmful and nobody gains anything as a result of it. In the pooling equilibrium the safe customers also suffer a negative externality by paying more per unit of cover and obtaining less cover than they would in a full information world or in the absence of the risky customers. In this case, however, the risky customers may be said to gain at the expense of the safe ones, since they obtain cheaper insurance and higher utility than they would under a full information world; the externality in this case, then, is not purely dissipative.

7.5 Recommended reading

The seminal work on the selection problem in insurance markets is Rothschild and Stiglitz (1976) and the reader is strongly recommended to look at it, since besides being a very important paper it is also very accessible. For an extension

7.6 Problem

Problem 7.1
Consider a competitive insurance market where individuals wish to insure their cars against theft. There are two types of individual, *Good* risks and *Bad* risks. Good risks face a probability of theft of 0.25 and Bad risks face a probability of theft of 0.75. All individuals have a utility function (for values of w greater than 0) of $U = 100 - 100/w$, where w is the individual's level of wealth. Each individual is endowed with a wealth level of 65, which includes 20 for the value of the individual's car.

(a) Show that the individuals are risk averse.
(b) Imagine that insurance companies are able to categorise customers accurately according to type. Show both algebraically and on a diagram the zero profit constraint in terms of w_{NT} and w_T for an insurance company offering insurance to its Good risk customers, where w_{NT} and w_T represent the individual's level of wealth in the no-theft state and the theft state respectively. Show diagrammatically that the Good risk customer would choose full insurance. Explain why this is so and calculate the insurance premium that the customer would pay and the resulting level of wealth in each state.
(c) Maintaining the assumption that insurance companies can categorise their customers accurately, show both algebraically and on a diagram the zero profit constraint in terms of w_{NT} and w_T for an insurance company offering insurance to its Bad risk customers, where w_{NT} and w_T represent the individual's level of wealth in the no-theft state and the theft state respectively. Calculate the insurance premium that the Bad risk customer would pay and the resulting level of wealth in each state.
(d) Now imagine that the insurance companies are unable to distinguish between Good and Bad risk customers. Assume that the proportions of customers of each type in the population are such that there is a Nash separating equilibrium in this market. Illustrate the Nash equilibrium using a state–space diagram, show the market average fair odds line and explain why the equilibrium is a Nash equilibrium. In what sense can it be said that the Good risk customers are suffering from a negative externality in this equilibrium?
(e) Set out the maximisation problem which could be used to determine the premium paid by, and the amount of compensation promised to, a Good risk customer in the separating equilibrium, being careful to specify the

appropriate zero-profit line and the self-selection constraint. Explain why it is possible to solve this problem by simultaneous solution of the zero-profit and self-selection constraints and, hence, show that the equilibrium value for w_{NT} for the Good customer is approximately 64.5, implying a total premium of 0.5 and compensation of 2.

Part III
The Labour Market: Education, Signalling, Screening and Efficiency Wages

Part III

CHAPTER 8

The Selection Problem and Education

8.1 Overview

In this section of the book we examine some implications of asymmetric information in the labour market. In this chapter we examine the selection problem by looking at a model where there are two types of worker, each with a different productivity level. Workers know their own productivity levels but employers are unable to observe them. This leads to problems, since the employer would be willing to pay higher wages to the *high* productivity workers than to the *low* productivity workers if they could be distinguished from one another. This problem is essentially similar to the selection problem in the insurance market of Chapter 7, and workers may use education levels to distinguish themselves in this market in a similar way to the use of deductibles for that purpose in the insurance market. Section 8.2 introduces the model, makes the distinction between signalling and screening, and looks at the screening case. Section 8.3 then examines the signalling case and argues that though it at first appears to differ strongly from the screening case, this difference is more apparent than real. Section 8.4 offers some general discussion of the model and, as usual, the final two sections of the chapter present some recommended reading and problems.

Chapter 9 will look at the hidden action problem in the labour market. This problem occurs where employers are unable to observe the effort levels of their employees and may have the consequence that they choose to pay wages above the market clearing level in order to encourage workers to work harder. The implication of paying wages above the market clearing level is, of course, the existence of involuntarily unemployed workers.

8.2 Education and screening

Imagine that there are two types of worker. The *high* productivity worker has a constant marginal product of h and the *low* productivity worker has a constant marginal product of 1, where h is greater than 1. Employers compete for workers by adjusting their wage offers until in equilibrium they make zero profits. Education, measured by y, may be acquired by the two types of worker at a cost per unit of a_h for the high productivity worker and of a_l for the low productivity worker, where a_l is greater then a_h. We assume, for simplicity, that education does not affect productivity and does not provide utility. Nevertheless, we shall see that since it is cheaper for high productivity workers to acquire education than it is for low productivity workers to do so, high productivity workers may choose to use education as a way of distinguishing themselves from low productivity workers and gaining higher wages.

The full information case

Consider first the case where there is full information and employers are able to distinguish the high productivity workers from the low productivity workers. In this case, the equilibrium is obviously where employers pay workers a wage equal to their marginal product and the level of education for each type of worker is zero (since it neither provides utility nor increases productivity). This case is illustrated in Figure 8.1.

Figure 8.1 plots wages on the vertical axis against education levels on the horizontal axis. The horizontal lines at h and 1 show the zero-profit lines for employers hiring high and low productivity workers respectively, employing a worker at points below these respective lines representing positive profits for the employers, since at such points the wage is below the marginal product for

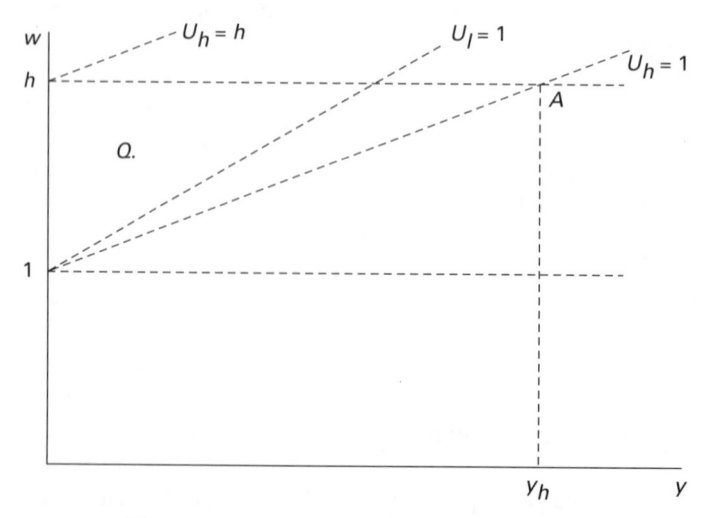

Figure 8.1 **The full information case**

the type of worker concerned. The upward-sloping lines represent indifference curves for the two types of worker as given by the following equation:

$$U_i = w - a_i y; \quad (i = h, l) \tag{8.1}$$

Equation (8.1) shows that workers are willing to acquire education in return for higher wages; since the cost of acquiring education is higher for low productivity workers, their indifference curves are steeper than those for high productivity workers – that is, low productivity workers need a bigger increase in wages to compensate for acquiring a given level of education than do high productivity workers. The indifference curves clearly satisfy the single-crossing property, and it is this property which makes the analysis of the model essentially similar to that of Chapter 7 despite other unimportant differences between the models. It does not matter for the analysis that the workers in this section are risk neutral, nor that they are not subject to uncertainty, and so differ from the risk averse individuals of Chapter 7 who were subject to the risk of having their cars stolen.

Under full information, employers will be able to pay different wages to the two types of worker, and competition drives them to locate on the respective break-even line for each type of worker. Since education serves no useful purpose in this model its equilibrium level is driven to zero. This is easily seen by considering, for example, that high productivity workers are offered the contract at point A in Figure 8.1 which specifies a wage of h provided the worker has an education level of y_h. Point A will clearly not be an equilibrium, since an employer could instead offer such workers the contract at Q with lower wages and lower education required. The high productivity workers would accept Q in preference to A, since the cut in wages is more than offset by the reduction in education costs, and places them on a higher indifference curve with a higher level of utility. Q will not, however, be an equilibrium, since it yields positive profits to the employers. Competition will drive employers to bid up wages and cut educational requirements in an attempt to attract more labour until the wage offer is h and the educational requirement is zero at the point on the break-even line for the high productivity workers which maximises the utility of the workers. Similarly, low productivity workers will locate, in equilibrium, at the point where w equals 1 and the educational requirement for them is also zero.

The analysis under full information is straightforward but it will serve as a useful benchmark to the analysis under asymmetric information, to which we now turn by assuming that employers are unable to distinguish between the two types of worker. This assumption may not always be appropriate but it may be reasonable in some cases – for example, where workers work in teams and it is difficult to observe the output of any individual worker. Anyway, it is the assumption that we shall employ, and we shall discuss the model later, in Section 8.4.

Screening versus signalling

Before proceeding with the analysis under asymmetric information it is useful to discuss the distinction between *screening* and *signalling*. Screening may be said to be the case where the uninformed party or principal, taking into account the asymmetry of information, designs the contract or contracts which he offers to the agents or informed parties before the agents take any action. The agents choose their actions after the principal has offered the contract or set of contracts to them. Signalling is the case where the agents choose their actions before the principal offers a contract or set of contracts.

The insurance model of Chapter 7 might, in terms of our new jargon, be termed a screening model, since it is natural in that case to think of the insurance company offering contracts and the insurees simply choosing over them. We shall call it a screening model even for the case where the equilibrium is pooling and the contracts do not screen the insurees into separate categories; that is, the important difference between screening and signalling models is in whether the agent takes any significant action before the principal makes his contract offers. Similarly, the selection problem in the market for investment funds in Chapter 2 may be considered to be a screening model, although we sometimes in the discussion allowed the entrepreneurs or agents to be offering contracts to the banks rather than the other way around.

Since it would not seem natural to imagine insurees offering different levels of deductibles to insurance companies before the companies offer contracts, the insurance market of Chapter 7 is considered to be a screening model. We could, however, imagine some actions other than the act of accepting a particular contract, such as installing a burglar alarm or engine immobiliser in a car, which might affect the cost of insurance and which might be considered to be an action taken by the insuree before the insurance contracts are offered, and so could be modelled as a signalling rather than a screening problem. Nevertheless, it seems intuitively easier to imagine the insurance market as a screening problem rather than a signalling one. At an intuitive level it seems easy, however, to imagine the selection problem in the labour market as either screening or signalling – that is, it seems easy to imagine workers choosing education levels in response to wage offers (screening) or to imagine employers making wage offers after workers have acquired education (signalling). We shall therefore analyse the selection problem in the labour market both as a screening problem and a signalling problem. We deal with the screening problem in the remainder of this section and the signalling problem in the next.

Education and screening

Now that the employer is unable to distinguish between the two types of workers, the full information equilibrium is no longer available; if employers were to offer wages of either h or 1 in this case, then both types of worker would accept the higher rather than the lower wage, and the employer would make losses. It may be possible, however, for the employer to vary the wage according to the educational attainment of the workers. Since the high productivity workers find it less costly than the low productivity workers to acquire education it may be the case that the wage can be increased for workers with a higher educational attainment in such a way that only the high productivity workers find it worthwhile to obtain the education, thus the two types can be separated and paid their appropriate wages. This analysis is directly analogous to adjusting the cost of insurance according to the size of the deductible in Chapter 7, where only the *safe* customers found it worthwhile to accept the deductible. Also, just as in Chapter 7, it is possible that the separating contracts may or may not produce a Nash equilibrium. The two possibilities are illustrated in Figure 8.2.

Figure 8.2 shows two cases, one where the separating contracts produce a Nash equilibrium and one where they do not. Consider first, Figure 8.2(i). The

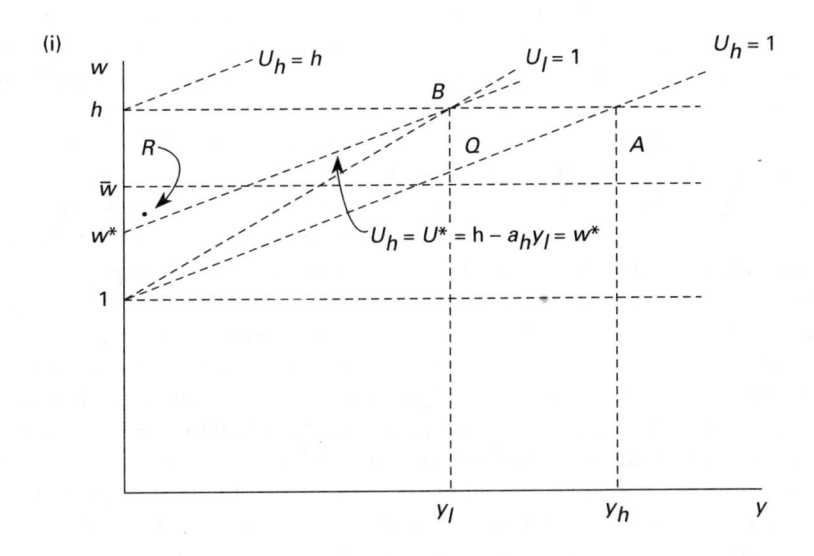

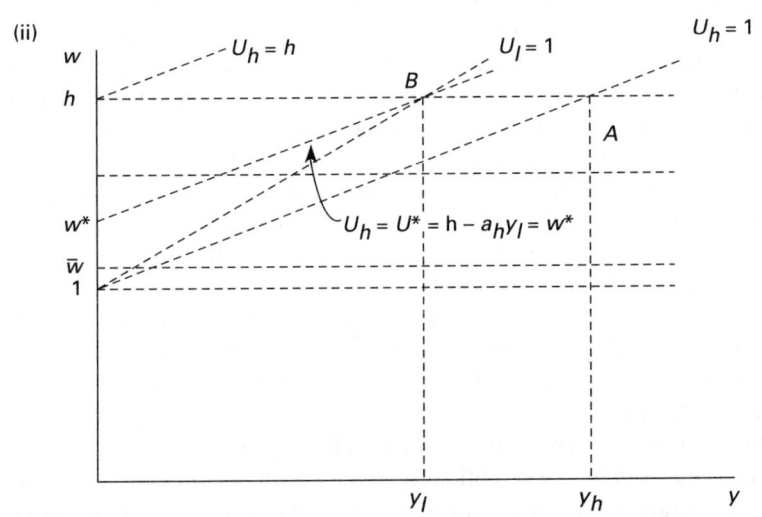

Figure 8.2 Separating contracts and Nash equilibrium

separating contracts that are candidates to produce a Nash equilibrium are those which offer a wage of 1 if the worker has zero education, or a wage of h if the worker has an education level of y_l; the low productivity workers will accept the former contract and the high productivity workers will accept the latter. This can easily be seen by starting with only the former contract on offer. Clearly, with only this one contract on offer the employers make supernormal profits since, assuming the participation constraints for both type of worker are satisfied and both accept the contract, employers are paying the high productivity workers less then their output at this wage. Thus competition will encourage employers to bid up wages.

Assuming for now that the employers bidding up wages wish to attract only the high productivity workers at the higher wages, it follows that the increased wages must be paid only to workers who have education levels sufficient to dissuade the low productivity workers from accepting the higher wages. The employers will achieve this goal as long as the wage–education combination that the new contracts offer produce a point on or below the indifference curve U_l equals 1 in the figure; such contracts will be less attractive to the low productivity workers than the contract specifying a wage of 1 for zero education. On the other hand, the new contract must be preferred by the high productivity workers. This requires that the contract produce a point above the indifference curve U_h equals 1 in the figure. The most education the high productivity workers could be induced to acquire would be y_h for a wage of h, which would locate them at point A in Figure 8.2, where they are indifferent between this point and the zero-education low-wage contract. Point A will not, however, be part of the separating set of contracts, since, if it were offered, an employer could deviate and offer the contract at point Q, which would both offer him profits and be preferred by the high productivity workers. Point Q will not be part of the equilibrium outcome either, since competition between employers will drive them to increase the wage offer and reduce the educational requirement until point B is reached. The contract at point B offers the low productivity workers the same level of utility as the zero education contract; as in the insurance case, we resort to the assumption of epsilon altruism to ensure that they do not choose the contract at B and impose losses on the employers.

The pair of separating contracts is found by the point of intersection between the indifference curve for the low productivity workers and the zero-profit line for the employer employing the high productivity workers in exactly the same way as the size of the deductible for the safe customers was calculated in Chapter 7. In other words, the separating contracts must, as for the insurance case, satisfy self-selection constraints. Competition between employers for high productivity workers results in the level of utility of U^* for high productivity workers; this is the highest level possible that is consistent with self-selection.

Having found the separating contracts, it is easy to see, again exactly parallel to the insurance case, the conditions necessary for them to produce a Nash equilibrium. Figure 8.2(i) shows the case where the separating contracts *do not* produce a Nash equilibrium. The reason for this is that a contract such as that indicated at point R in the figure will be preferable, for both types of worker, to the separating contracts and will yield an employer offering it supernormal profits. The reason why this contract offers supernormal profits is that it lies below the market average zero-profit line, which is the horizontal line at $\bar{w}$, where $\bar{w}$ is the average marginal product of the two types of worker weighted by their proportions in the population. Thus, at R, the wages are below the average product of the workers, and the employer makes supernormal profits.

The existence of a point such as R depends on the indifference curve for the high productivity workers through their separating contract cutting the market-average zero-profit line to the right of the vertical axis, which in turn depends upon the average productivity of the two types of worker and the slope of the indifference curves of the high productivity workers – that is, a_h. Figure 8.2(ii) draws the case where the separating contracts *do* produce a Nash equilibrium, since the indifference curve for the high productivity workers through their separating contract in this case does not cut the market-average zero-profit line to the right of the vertical axis, and no profitable deviation from the separating contracts can be found.

Formally, the separating contracts will produce a Nash equilibrium if the following condition holds:

$$U_h(h, y_l) = h - a_h y_l \geq U_h(\bar{w}, 0) = \bar{w} \tag{8.2}$$

where $U_h(h, y_l)$ is the level of utility level gained by a high productivity worker accepting the separating contract designed for him, and $U_h(\bar{w}, 0)$ is the utility he would earn if offered a wage equal to the market-average productivity level and acquired no education (that is, the terms in the brackets represent the wage-education combination specified by the zero education contract).

Just as in the insurance case of Chapter 7, no pooling contract can ever produce a Nash equilibrium (as the reader is asked to verify in problem 8.1 below). Hence, when the separating contracts do not produce a Nash equilibrium, it is necessary to consider alternative equilibrium concepts and, as for the insurance case, it may be shown that the Wilson equilibrium is the pooling contract (with wages of $\bar{w}$ and education levels of zero) and the Reactive equilibrium is the pair of separating contracts (the reader is asked to verify this in problem 8.2).

Before going on to the signalling case in the next section, notice that, once more similar to the insurance case where the safe customers suffered a negative externality from the presence of the risky customers, the high productivity workers suffer from the presence of the low productivity workers when asymmetric information creates the selection problem. In the separating equilibrium there is a dissipative externality, the high productivity workers being worse off than in the full information case (since they have to acquire costly education to distinguish themselves from the low productivity workers, and earn the appropriate wages of h while the low productivity workers do not gain anything and locate at their full information wage–education combination). In the pooling equilibrium the high productivity workers again suffer, this time from receiving a wage equal only to the average level of productivity, but the low productivity workers gain from receiving such a wage. In each case, however, the equilibrium is efficient, given the information constraint – that is, given the informational asymmetry it is impossible to find a way of achieving a Pareto improvement.

Although from a social point of view the expenditure on education by the high productivity workers in the separating equilibrium may be viewed as waste, since it is neither productive nor gives utility, it is not viewed as wasteful expenditure by the high productivity workers, since it allows them to distinguish themselves from the low productivity workers and increase their welfare by gaining an increase in wages which more than offsets the loss due to expenditure on education. Education is able to serve this function because it is cheaper to acquire for the high productivity workers, thus producing the relative slopes of the indifference curves of the two types of worker and the single-crossing property. The role of education here is directly analogous to the role of deductibles in Chapter 7, so the reader who has found this section difficult should study the previous chapter once more before re-examining this section.

8.3 Education and signalling

We now consider the case where workers acquire education before employers offer contracts. In this case, the beliefs of the employers concerning how to interpret the education signal become crucial for determining the resultant equilibrium. Arbitrarily defining these beliefs leads to a multiplicity of possible equilibria but we shall see that it is possible to argue that some beliefs make more sense than others and to make the signalling case effectively produce the same equilibrium outcomes as the screening case.

Given the importance of the employers' beliefs in the signalling case we need to reconsider the concept of equilibrium. The extra element we need to consider is that the beliefs of the employers in equilibrium must not be inconsistent with the evidence given to the employers. If they are, the evidence will be used to adjust beliefs. Since the process of adjusting beliefs in response to evidence is usually associated with Bayesian statistics and carried out using Bayes' Rule, the new equilibrium concept is known as a *perfect Bayesian equilibrium*. We do not need to enter into the statistical or game theoretic details here, it will suffice to illustrate the ideas using the signalling version of the model of the previous section and Figure 8.3.

Begin by specifying arbitrarily that employers believe that any worker seeking employment who has an education level of y^* or more is a high productivity worker and so will be paid h, and that anyone with an education level less than y^* is a low productivity worker and so will be paid 1. If employees' actions, given these beliefs, lead to evidence which does not refute them then we shall have a perfect Bayesian equilibrium.

The Selection Problem and Education 123

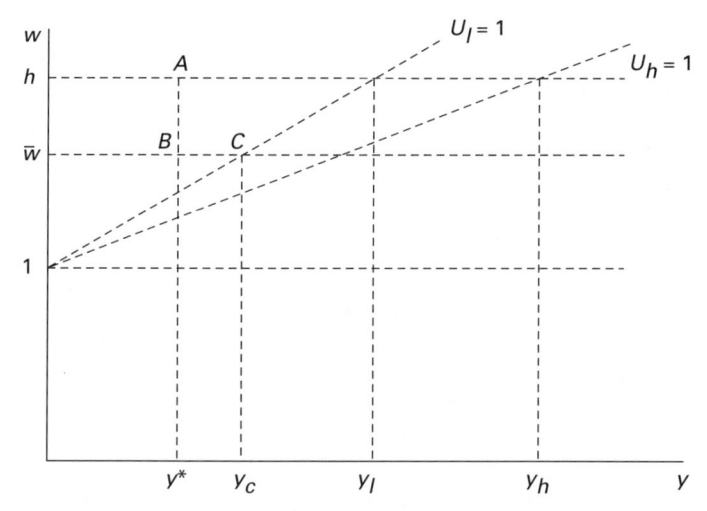

Figure 8.3 Signalling and education

In Figure 8.3, y^* is shown as a value between zero and y_l. Given the beliefs of employers, both types of worker would clearly prefer to acquire education to the level of y^* and earn wages of h at point A in the figure rather than to locate at the wage of 1 on the vertical axis. Workers of either type would locate on higher indifference curves if they acquired an education level of y^* than if they acquired no education. Workers optimal choices, given the employers' beliefs, thus lead to evidence which refutes those beliefs, since workers with an education level of y^* are not all high productivity workers as specified in the employers' beliefs. Indeed, employers acting on those beliefs will hire a random sample of workers of mixed ability and make losses by paying all of them the wage for a high productivity worker. Thus the beliefs which we specified do not lead to a perfect Bayesian equilibrium.

It is easy to see that a perfect Bayesian equilibrium could be produced if employers believed that workers with an education level of y^* or more were likely to be high productivity workers with probability p, and low productivity workers with probability $(1-p)$, where p is the proportion of high productivity workers in the population. In this case, employers would be willing to pay the wage $\bar{w}$ equal to the average productivity of the workers for workers with education levels of y^* or more. Workers with lower levels of education are believed to be low productivity workers and would be paid 1. Workers of either type again find it optimal to send the information signal of y^* to the employers, but employers now expect to break even by hiring a random sample of workers, and their beliefs are consistent with the evidence they receive: that is, the proportion, p, of workers hired will be high productivity workers, just as the employers expected. Thus the beliefs produce a perfect Bayesian equilibrium.

There are, however, problems with the perfect Bayesian equilibrium at point B in Figure 8.3. Notice that all workers acquire the education signal, which serves no useful purpose and should properly be regarded as wasteful. Holding everything else the same in the last paragraph, but replacing y^* by $y^*/2$ would produce a better pooling equilibrium with only half as much waste on education. Indeed, the best pooling equilibrium of all is that where an education level of zero signals that workers are likely to be high productivity workers with probability p and low productivity workers with probability $(1-p)$; this equilibrium would also produce a perfect Bayesian equilibrium. Thus there are many possible perfect Bayesian pooling equilibria which may be produced by replacing y^* in our discussion by any value between zero and y_c shown in Figure 8.3. It is necessary, therefore, to devise a way of choosing between the large number of possible equilibria. Fortunately this task is not too difficult, but before undertaking it let us look at the possibility of separating equilibria.

It is easy to see that the separating contracts of the previous section can also be a basis for a separating equilibrium under signalling. If employers' beliefs are that anyone with an education level of y_l or more is a high productivity worker and will be paid h, while anyone with a lower level of education is a low productivity worker and will be paid 1, then their beliefs will not be refuted by the evidence. Given the employers' beliefs and wage offers, the low productivity workers will choose zero education and low wages, while the high productivity workers will choose the higher levels of education and wages. Thus the equilibrium will be a perfect Bayesian equilibrium. However, just as there are many possible pooling equilibria in the signalling model, there are many possible separating equilibria too. It is possible to replace y_l in the above separating equilibrium by any value for y between y_l and y_h without otherwise affecting the analysis. For values of y above y_h the story becomes different, since even the high productivity workers then choose zero education, forcing the employers to revise their beliefs about how to interpret a signal of zero education. On the other hand, for values of y below y_l the story becomes different because then even the low productivity workers would choose to acquire education.

Thus there are many possible equilibria in the signalling model, each of which represents a possible perfect Bayesian equilibrium dependent upon a particular set of beliefs held by the employers and with different levels of education being chosen. This represents a rather unsatisfactory state of affairs and it is desirable to narrow down the number of possible equilibria, preferably to a unique equilibrium. This can be done by using the communication tests of Hirshleifer and Riley (1992).[1]

Consider first the set of separating equilibria. Of all of these it is possible to show that the only one to avoid elimination is the efficient separating equilibrium – that is, the one with the least expenditure on education by the high productivity individuals or the one that is identical in outcome to the separating contracts found in the discussion of screening. In the jargon of Hirshleifer and Riley, only the separating equilibrium with a level of education for the high productivity workers of y_l is *weakly communication proof*.

The Selection Problem and Education   125

An equilibrium is weakly communication proof if no message or communication from a worker to an employer of the following sort is credible: 'I am a worker of a certain type. You should believe me, for if you do and respond optimally I will be better off, while if I were of any other type I would be worse off.' For example, consider a separating equilibrium where employers believe that workers with an education level of higher than y_l are high productivity workers. Such an equilibrium would not be weakly communication proof since a high productivity worker could choose y_l and then credibly say to the employer: 'I am a high productivity worker. Believe me and pay me h because it would not be worthwhile for a low productivity worker to have acquired education of y_l in order to improve his wages from 1 to h.' Thus, of all the separating equilibria, only the one with an education level of y_l for the high productivity workers will pass the weak communication test.

However, just as in the screening case, it was necessary to consider whether the separating contracts produced a Nash equilibrium, here it is necessary to see if the weakly communication-proof separating equilibrium is also *strongly communication proof*. It will be so only if no message of the following sort is credible: 'Both types of worker would prefer to be paid a wage equal to the average productivity, $\bar{w}$, rather than to accept the separating contracts. Therefore, although I have chosen zero education rather than an education level of y_l you should use the population proportions in estimating my worth and so pay me $\bar{w}$.' No message of this sort will be possible if the indifference curve for the high productivity workers through their separating contract does not cut the market-average zero-profit line to the right of the vertical axis – that is, the separating equilibrium under signalling will be strongly communication proof under exactly the same conditions required for the separating contracts under screening to provide a Nash equilibrium; that is, when constraint (8.2) holds.

The difference between screening and signalling may therefore be more apparent than real. If screening produces a Nash separating equilibrium, then under signalling there will be a strongly communication proof equilibrium with exactly the same wage–education combinations and choices as in the screening Nash equilibrium. If no Nash equilibrium exists under screening then no strongly communication proof equilibrium will exist under signalling. Furthermore, just as no pooling contract produces a Nash equilibrium under screening it may be shown that no pooling contract ever produces a weakly communication proof equilibrium under signalling (as the reader will be asked to verify in problem 8.3).

8.4 Discussion

Education in the above analysis performs a useful role in a separating equilibrium as a means for workers to transmit useful information about their type to employers (a role it can play because of the different costs of acquiring education for the two types of worker) but it has no useful role either as a way of improving workers' productivity or as a way of providing utility. It is, however, possible to allow for such factors (which one hopes makes the analysis more realistic!) without changing the essence of the analysis. For example, when education improves worker productivity it is still the case that there will be more resources expended on education in a world of asymmetric information than in a full-information world, and this extra spending may be viewed as waste (apart from its value in distinguishing between workers) rather than all the spending on education being waste in our analysis.

Although the screening/signalling role for education seems (to me at least) to be intuitively appealing, it is possible to argue that the empirical support for it is weak. For example, Layard and Psacharopoulos (1974) found that college dropouts get as high a return on education as those who gain degrees. This result, however, might be consistent with years of education being the signal rather than the type of qualification gained. Alternatively, if there are diminishing returns to time spent studying it may be that dropouts decide rationally that it is not worth their while to study for as long as those who continue in education if it is the case that some people have different time-return profiles for education than others and dropouts simply reach the optimal return rate sooner than the rest. They also found that tests, which may be a cheaper way of distinguishing between types of worker than education, are not widely used, but this may be explained if education has other useful roles to play, in providing utility or improving productivity, in which case one might not so easily expect testing to replace education. Also, the rate of return to education appears to increase over time in the sense that wage differentials between people with different education levels increase over the years after education has been completed. This last result may seem to conflict with education playing its signalling/screening role in the early years after education (when qualifications would seem to be most important in helping someone to find employment) but, on the other hand, if workers are categorised by education and those workers with higher education levels are then given jobs which require more on-the-job learning (because they have shown their ability to learn) it would follow that their wages rise as they learn how to do their jobs better over time. Thus empirical results need to be interpreted carefully and it is difficult to carry out straightforward and meaningful tests without allowing for relaxation of the simplifying assumptions of a theoretical model.

8.5 Recommended reading

The seminal article on education as a signal is by Spence (1973) and is well worth reading.

8.6 Problems

Problem 8.1
Consider the efficient pooling contract for the screening model depicted in Figure 8.2(i). Under this contract the wage offered is $\bar{w}$ for an education level of zero. Explain why this contract, or any other pooling contract in this model, cannot produce a Nash equilibrium.

Problem 8.2
Consider the screening model depicted in Figure 8.2(i).

(a) Show that for this model there is no Nash equilibrium under separating contracts.
(b) There is no Nash equilibrium under either pooling or separating contracts for this model. Explain what is meant by the Wilson equilibrium and apply it to the model in Figure 8.2(i).
(c) Explain what is meant by the Reactive equilibrium and apply it to the model in Figure 8.2(i).

Problem 8.3
Consider the efficient pooling contract for the signalling version of the model depicted in Figure 8.2(i). Under this contract, the wage offered is $\bar{w}$ for an education level of zero. Explain why this contract, or any other pooling contract in this model, cannot produce a weakly communication proof equilibrium.

CHAPTER 9

The Hidden Action Problem and Efficiency Wages

9.1 Overview

One of the most important issues in economics is that of unemployment. Why does the labour market fail to clear? A potential answer to this question is provided by the idea of efficiency wages.[1]

There are several variants of efficiency wage models. The common element in all of them is the idea that the quality of the workforce employed by a firm depends on the wages paid by the firm in such a way that reducing the real wage would reduce the quality of the workforce and reduce the profits of the firm if pushed too far. Hence employers have a reason not to cut wages even if it would be possible to force wage cuts on the existing labour force or to hire cheaper workers from the pool of unemployed workers. Thus the wage may not fall to its market clearing level.

The failure of the real wage to clear the labour market is not, in the efficiency wage analysis, a result of any rigidity such as might be imposed by trade union or worker reluctance to accept wage cuts or government imposed minimum wages, nor is it a result of some Keynesian-style demand failure; rather, it is simply a result of private optimising behaviour.

Several reasons may be put forward to explain why it benefits a firm to pay higher wages. We shall review them briefly in the following section before going on to look in detail in section 9.3 at one specific model based on the hidden action problem. Section 9.4 will recommend some reading and Section 9.5 presents some problems to check the reader's understanding.

9.2 Reasons for paying efficiency wages

Nutrition

The efficiency wage hypothesis was first put forward in relation to less developed economies where there may be people willing to work for 'almost nothing'. However, if a worker earns 'almost nothing', he may be so weak as to produce 'almost nothing'. In order to provide a decent day's work, workers need to have the energy to do so and it pays firms to pay workers more than the minimum wage needed to hire them so that they will be able to afford more food and be more productive. Similar arguments may be applied to the interesting evidence that Percival Perry, the manager of a Ford motor car plant in Manchester in the early part of the twentieth century, observed that a typical male worker needed weekly wages of £3 in order to provide for himself and his family compared to the actual wage paid of only £1.50. Consequently, Perry increased wages to £3 per week and reaped considerable productivity gains.

Absenteeism and labour turnover

If workers quit their jobs and need to be replaced this may be costly for the employer, since there may be a loss of output if it takes time to hire or to train new workers. Hence firms have an incentive to pay higher than market clearing wages in order to reduce labour turnover, since the higher the wages the less likely it is that workers will wish to quit.

The above reasoning may explain a famous action by Henry Ford. Ford approximately doubled the wages of much of the workforce in his Detroit plant in 1914 and introduced a $5 day. He did this despite the fact that there was no shortage of applicants for jobs at the initially lower wages and there was a great surplus of applicants at the new higher levels of pay. Nevertheless, Henry Ford described the pay rise as 'one of the finest cost cutting moves we ever made', which would indicate that, in terms of our jargon, he thought of the pay rise as moving pay towards the level of efficiency wages.[2]

The reasons why it suited Ford to pay higher wages appear to be that they were necessary to induce the workers to tolerate the boring and repetitive assembly-line production techniques being pioneered by Ford. The payment of higher wages reduced absenteeism and turnover, both of which may have been very costly for the Ford plant, and so proved to increase the quality of the workforce (in the sense of better morale, lower absenteeism and turnover). Of course, since the tasks Ford wished his workers to carry out were mainly simple and repetitive, the training cost element stressed in some discussions of turnover models was not so important in explaining the move to the $5 day. Nor was it necessary for most of Ford's workers to be highly skilled or able (indeed, to some extent he wanted untrained workers who would more easily accept his new production line regime) so that it is unlikely that his decision to increase pay was much motivated by our next possible reason for efficiency wages, the selection problem.

The selection problem

This variant of the efficiency wage model depends on the joint hypothesis that better-quality workers are less willing to take a job for low pay than poorer-quality workers, and on the inability of the employer to distinguish between types of worker. Thus paying lower wages might attract only poor workers, whose productivity differential is even greater than their wage differential, so that it is not worthwhile for the employer to pay lower wages. One reason why high-quality workers might prefer to remain unemployed rather than take a job on low pay is that taking such a job might be interpreted by potential employers as a signal either of low quality or low reservation wage (in the latter case the employer might feel that he can attract the worker to take a job while paying him only marginally more than in his current low-paid job).

It is often argued that this model may help to explain racial or sexual discrimination in the workplace. The argument is that if some group of workers is known to be willing to work for lower wages than some other group, then employers will use this knowledge to pay them less than the others. This does not, however, seem to me to be a convincing variant of the selection problem, since the argument is based not upon unobservable characteristics but upon observable ones and the use of uneven market power. The implication of the argument is that the workers discriminated against provide cheaper-than-average labour for employers. If this is so, then one might expect competition to hire such workers to raise their wages to the average level over time.

Sociological models

Akerlof (1982) has argued that enduring work relationships between employers and employees cannot be explained without some reference to group behaviour and work norms. According to this argument, it might not be worthwhile to try to pay each worker according to his or her marginal productivity, since such payment mechanisms would be costly to operate and, in any case, worker morale, flexibility and productivity might profitably be improved by paying higher than market clearing wages to maintain good employer–employee relations. Such models are called 'gift-exchange' models, the idea being that if the employer is generous to his workers, they will return the favour by being flexible and hard-working.

The hidden action or shirking model

This model is based on the idea that it may be difficult to observe the work effort of individual employees even though productivity and output rise with effort.

Therefore it may be in a firm's best interest to engage in a limited monitoring activity to observe work effort while paying above market clearing real wages. Any worker caught being lazy or shirking is fired, so a worker has an incentive to work hard in order to keep his job rather than to be caught shirking and face the punishment of a period of unemployment. In equilibrium all firms pay higher than market clearing real wages. No firm has an incentive to reduce wages despite the existence of unemployment, since if it cut wages then its workers would not fear being fired and would therefore shirk and so reduce the firm's output and profits.

Notice that with the wages set at their efficiency levels, workers in this model do not decide to shirk. Thus the unemployed are not made up of shirkers who got caught and fired but are those who decide to quit their jobs, or are new entrants to the labour market, or those made redundant by firms facing difficulties.

We shall examine this model in detail in the next section and see that the optimum or efficiency wage will depend on the wages paid by other firms, the level of unemployment and the level of employment benefit, all of which affect the size of the threat of dismissal.

Dual or multiple labour markets

The models outlined above may explain why the labour market is divided into different submarkets. For example, if shirking is a problem in some sector or firm where monitoring of work effort is costly and shirking expensive, then we might expect to observe the payment of efficiency wages in this sector or firm but not in others, known as the *secondary sector*, where perhaps monitoring is not such a problem. The market clearing wage may prevail in the secondary sector. Unemployment might be viewed as either voluntary, in the sense that a job could be found at some wage in the secondary sector, or involuntary, where the unemployed worker would be willing to accept a job in the efficiency wage sector if he were offered one.

Discussion

The above models are not immune to criticism. For example, if a threat is needed to discourage shirking behaviour why must it take the form of unemployment and loss of an above market clearing real wage? Why can't a worker pay a fee or post a bond redeemable to his employer if he is caught shirking? This criticism does not seem to recognise that the workers concerned may have no savings or assets with which to pay such a fee, and the scheme is potentially open to abuse by employers who unjustly claim that a worker was shirking in order to fire him and claim the bond.

Another alternative which may seem more plausible would be for an employer to pay seniority wages; that is, wages which rise the longer the employee remains employed. The employee has an incentive to keep his job and reap the gains as he moves up the seniority ladder. A possible problem with this scheme is that the firm has an incentive to hire workers on a low wage and promise seniority pay rises but then to renege on this promise and fire the workers as they become senior. Nevertheless, it is possible for firms to acquire a reputation for honesty and to use seniority pay in this way.[3]

Seniority pay schemes or the posting of bonds may be viewed as a response to the hidden action problem where the threat used to prevent shirking does not require the existence of a pool of unemployed workers as in the more basic theory where the wages are increased above the market clearing level. Similarly, the adverse selection problem might be tackled by using aptitude tests or interviews to try to select workers rather than paying above market clearing real wages, while, if training costs are important, the firm might be able to respond by paying less to untrained than to trained workers, or by charging workers a training fee (which would, of course, only work if the worker had the funds to pay the fee). All these points seem to have some merit; we do observe seniority scales, testing and training fees, but for the efficiency wage theories to have relevance for explaining unemployment simply requires the existence of a residual problem even after all responses such as seniority wages have been implemented. However, even if efficiency wage theories do have a role to play in explaining unemployment it should be noted that they are usually thought to explain equilibrium levels of unemployment and not cyclical variations in unemployment.[4] Also notice that efficiency wage theories do not appear to call for Keynesian-style intervention via boosts to aggregate demand, since they are theories of real wage adjustment in response to informational problems and not theories of money wage rigidities or demand shortfalls.

9.3 The hidden action problem and the shirking model

Consider a simple model where there are N identical workers who live for ever and maximise their lifetime utility by choosing whether to apply for jobs and whether to work or shirk if employed. If employed a worker's utility is given by:

$$U = w - e \tag{9.1}$$

where w is the real wage and e is the disutility from effort expended, which may, for simplicity, take two levels, either 1 or 0 (work or shirk).

The employed worker will choose to work rather than shirk only if the *no-shirking constraint* is satisfied; that is, if expected utility from working is greater than or equal to the expected utility from shirking. It is possible to show (see Shapiro and Stiglitz, 1984) that the no-shirking constraint may be written as:

$$w - 1 \geq d + (a + b + r)/q \tag{9.2}$$

where d is the level of unemployment benefit received by an unemployed worker, a is the probability per unit time of an unemployed worker receiving a job offer, b is the probability per unit time of an employed worker being made redundant as a result not of being caught shirking but of his firm facing difficulties, r is the worker's rate of time preference, and q is the detection rate or the probability per unit time of a shirking worker being caught shirking. The interpretation of constraint (9.2) is straightforward. Obviously, the no-shirking constraint will only be satisfied if the wage minus the unitary disutility from working more than compensates for the level of unemployment benefit, d. By exactly how much $w - 1$ must exceed d depends upon a number of obvious factors. The wage necessary to satisfy the no-shirking constraint must rise with a, since an increase in a reduces the threat of unemployment. Similarly, the wage must rise with b since a higher chance of losing one's job through no fault of one's own makes the job less worth working hard to keep. Also, a rise in the rate of time preference, r, means that more weight is attached to the present gain from shirking than from the future gains of wages from continued employment. Finally, notice that the wage necessary to satisfy the no-shirking constraint falls as the detection rate, q, rises, since if the chance of being caught shirking is high then the wage to induce working rather than shirking need not be so high. We shall treat d, a, b, and r as given, but in a fuller model it would seem that d, a, and b, should be determined endogenously.

The no-shirking constraint is illustrated in Figure 9.1. The figure plots real wages, w, on the vertical axis against aggregate employment, L, on the horizontal axis. If the employers were able freely to observe shirking behaviour, so that q took the value of unity, then the supply of non-shirking labour function would be the reverse-L-shaped function shown as L_S, indicating that workers would be willing to accept employment for any wage above $d + 1$ to compensate for unemployment benefit and the disutility of effort. However, given the hidden action problem, workers will accept a job and work rather than shirk only if the

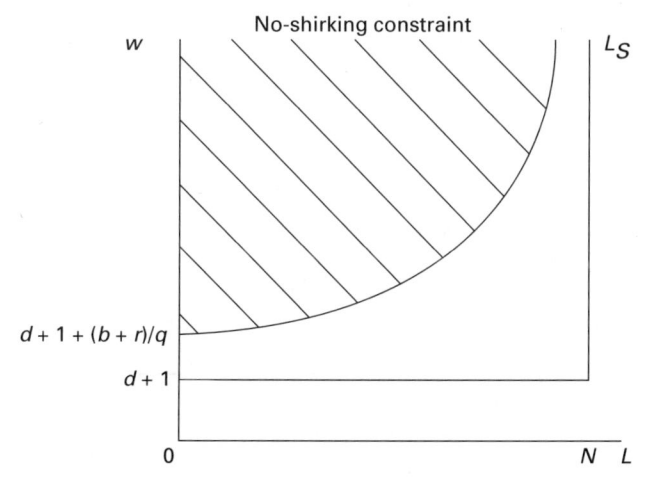

Figure 9.1 **The no-shirking constraint**

no-shirking constraint is satisfied; that is, only if the wage lies in the shaded region in the figure. The boundary between the shaded and non-shaded regions is the locus of points where the no shirking constraint is met with equality, and points within the shaded region are points where the constraint is met with inequality.

The shape of the no-shirking constraint may be explained as follows. If L is zero, the probability of gaining employment if unemployed, a, is zero, so the minimum wage necessary to satisfy equation (9.1) is $d + 1 + (b + r)/q$, hence the no-shirking constraint cuts the vertical axis at this wage. As L rises beyond zero it causes a to rise and this in turn causes the minimum wage necessary to satisfy the no-shirking constraint to rise, as shown.

Now assume, for simplicity, that unemployment benefits, d, are set to zero. In this case an employer's only cost of hiring a worker is the wage paid. Also assume diminishing marginal productivity of employment so that the aggregate demand for labour may be plotted as the downward sloping curve L_D in Figure 9.2. It is assumed that the worker's productivity is zero if he shirks, so we may assume that employers are only interested in hiring workers who decide not to shirk.

If shirking were costless to observe, then employment and real wages would be given at the intersection of the L_S and L_D lines at point A in Figure 9.2. In the presence of the hidden action problem, the equilibrium occurs at point B, since at point A the no-shirking constraint is not satisfied. Each individual firm will treat a as given independently of its own actions and hire workers until the demand for labour line cuts the no-shirking constraint at point B.

Employment, at L^*, will be below the full information level and there will be involuntary unemployment of $N - L^*$ in equilibrium. Wages, at w^*, will be above the full information level. Even though the unemployed would accept jobs and

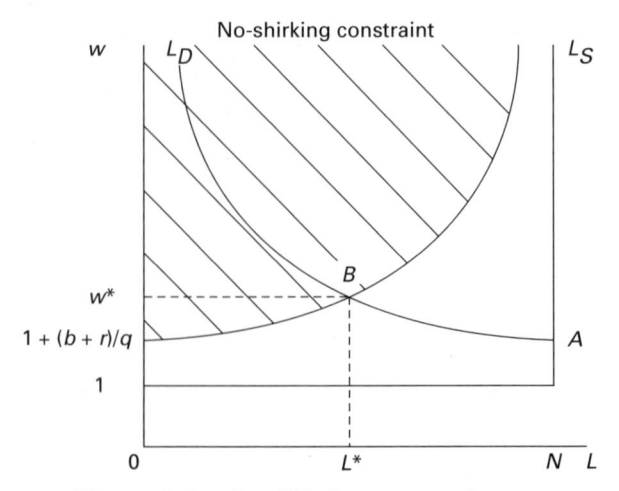

Figure 9.2 Equilibrium unemployment

be willing to work rather than shirk for wages below w^*, no firm will wish to hire them at such wages, because once hired the workers would then choose to shirk. If the unemployed could commit themselves to work rather than to shirk they could obtain employment but, in a manner similar to the insuree unable to commit himself to take care rather than to be careless, or an entrepreneur unable to commit to investing in a certain project, the worker cannot make such a commitment credible and suffers as a result. Notice that if possible a movement from B to A would be welcomed by both unemployed workers and employers, but would be to the detriment of the workers actually employed at B and would not therefore represent a Pareto improvement.

9.4 Recommended reading

A useful survey of the literature is provided by Yellen (1984). The model of the previous section is based on Shapiro and Stiglitz (1984). Akerlof (1982) provides the seminal reference on the sociological model. The paper by Raff and Summers (1987) asks the question, 'Did Henry Ford pay efficiency wages?' and provides a fascinating and very enjoyable examination of the evidence, including the earlier episode when Perry raised wages at Ford's Manchester plant.

9.5 Problems

Problem 9.1
'All efficiency wage models are based upon the idea that the quality of the workforce employed by a firm depends on the wages paid by the firm in such a way that reducing the real wage would reduce the quality of the workforce and reduce the profits of the firm if pushed too far.' Discuss.

Problem 9.2
Consider the shirking model illustrated in Figure 9.2. What, if any, would be the effects of a rise in the detection rate, q, on the position of the no shirking constraint and the levels of real wages and employment in the presence of the hidden action problem?

Part IV
Regulation, Public Procurement and Auctions

Part IV

CHAPTER 10

Regulation and Procurement

10.1 Overview

Governments in many countries regulate the behaviour of private companies that are considered to have some element of monopoly power. Obvious examples of such companies are the so-called public utilities such as telecommunications, gas, electricity and water.[1] The aim of regulation is to prevent the abuse of monopoly power and encourage behaviour consistent with the welfare goals of the government. Since the managers of the regulated companies know more about their companies than do the regulators, it is important to take this asymmetry of information into account when designing regulatory policy. Similarly, when a public-sector body wishes to commission a large-scale public project, such as a new telecommunications system or the building of a new port, from a private sector company (or group of companies), the purchasing authority may be unaware of the private sector company's cost structure and face an asymmetric information problem in fixing a price for the project.

We shall look in section 10.2, at a highly simplified version of the regulatory problem with hidden information, and then look in section 10.3 at the procurement or purchasing problem with hidden actions and hidden information. Section 10.4 suggests some further reading, and section 10.5 presents a problem.

10.2 Regulation and hidden information

Consider the case of the regulator of a private monopolist who wishes to maximise welfare, W, defined as:

$$W = S + \Pi \tag{10.1}$$

where S represents consumer surplus of the customers of the monopolist, and Π represents the profits of the monopolist. The monopolist has a very simple production function with constant average and marginal costs equal to C. The regulator knows the demand schedule of the consumers, but knowledge of the production function facing the monopolist is the private information of the monopolist. The regulator is unable to observe unit costs, either *ex ante* or *ex post*, but knows that they will take either the value C_H or C_L; where C_H exceeds C_L and the regulator attaches the probability of p to the true value of C being C_H. The regulator exercises control by setting the price of the output of the monopolist. The price must be chosen to satisfy the participation constraint of the monopolist, which requires that profits must be greater than or equal to zero.

Under full information, the regulator's task is quite simple: price should be set equal to unit cost C. This is illustrated in Figure 10.1.

Figure 10.1 shows the demand curve for the product of the monopolist, D, as well as the unit cost function, C. The regulator will maximise W if he sets price, P, equal to the unit cost C. Under this pricing rule, output will be set equal to demand at Q_1 and W will be equal to the consumer surplus represented by the shaded areas X, Y and Z (which represent the amount of surplus gained by the consumers, who would be willing to pay a price in excess of C for quantities less than Q_1).[2] The monopolist will simply cover his costs (which total $V + W$) at this price level, and Π will be zero.

It would be possible to make Π positive by increasing the price level, for example to P_1, but this would increase profits by only the shaded area Y (representing the excess of total revenues, $V + Y$, over total costs, V, at the new price and output levels) while causing consumer surplus to fall by the areas Y and Z.

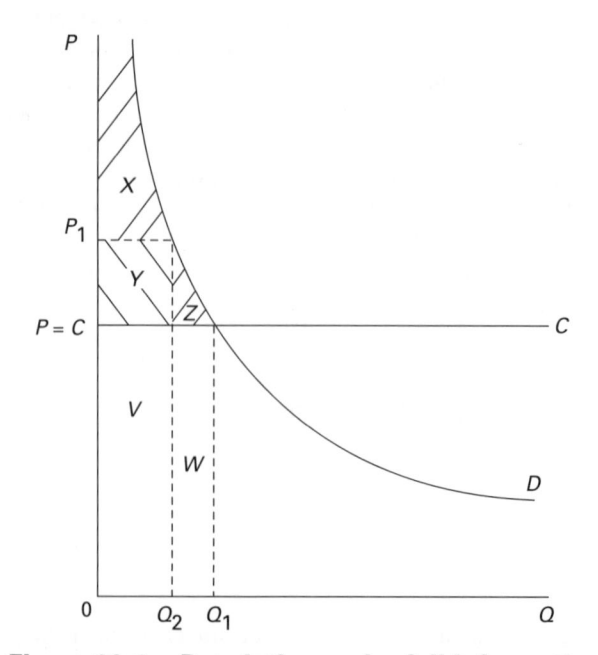

Figure 10.1   Regulation under full information

Hence, raising the price to P_1 would lower social welfare, W, by the amount represented by area Z.[3] The regulator may not lower the price below C, since a lower price would not allow the monopolist to satisfy his participation constraint (and would, in any case, lower W, as the reader may check, even if the monopolist was prepared to operate at a loss).

Thus, under full information, the regulator's optimal policy is just to set price equal to the constant average and marginal cost of the output. However, he can no longer simply follow this rule when he is unsure of the monopolist's unit costs. This case is illustrated in Figure 10.2.

Figure 10.2 shows the two potential unit cost functions which the regulator believes the monopolist may have. If the regulator knew the value of unit costs he could set price equal to unit cost, either C_H or C_L, leading to output of either Q_H or Q_L as shown. If we assume that the regulator wishes to ensure that some production takes place, then he must set a price equal to the higher value of C_H when he does not know the value of unit costs. This is so, since a monopolist with low costs will not wish to reveal that information to the regulator and if offered a choice of prices would always choose a price above C_L, while at any price below C_H the participation constraint of the monopolist with high costs will not be satisfied and he will cease to produce if he faces a price less than C_H. This means that a low-cost monopolist will make supernormal profits, illustrated by the shaded area Y in Figure 10.2, and that the sum of welfare when the regulator faces a low-cost producer will be given by $X + Y$ rather than by $X + Y + Z$, which could be achieved if the regulator knew he was facing a low-cost producer. The net loss of welfare, Z, when the regulator faces a low-cost producer, means that there is an expected cost resulting from the asymmetry of information of pZ (since the probability of facing a low-cost producer is p).

We have, of course, assumed in the above analysis that the difference in possible costs of $C_H - C_L$ is not so great that the price of C_H would be higher than

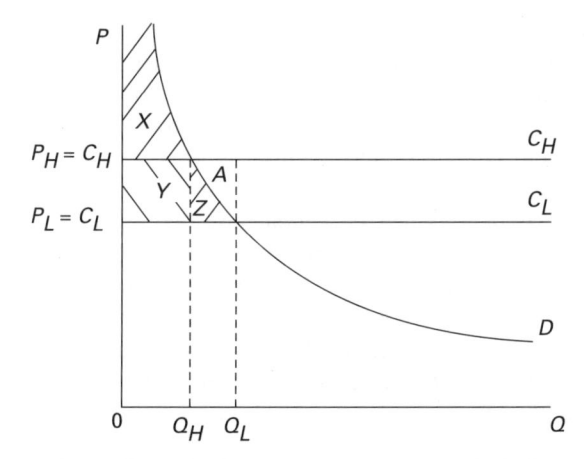

Figure 10.2 Regulation under asymmetric information

that which would be chosen by the low-cost producer if he was able to act as an unregulated monopolist; if this were the case, then the regulator would do better simply to allow the monopolist to set his price without regulation.

The regulator can avoid the expected cost of pZ caused by the asymmetry of information if he is able to offer a subsidy, T, to the monopolist as well as specifying the price that the monopolist must charge for his product. The optimal regulation in this case is to offer the monopolist a choice of contracts, as follows:

$$T = X \text{ if } P = C_H \tag{10.2a}$$
$$T = X + Y + Z \text{ if } P = C_L \tag{10.2b}$$

The monopolist must agree to satisfy demand at the price level specified by the contract he chooses. The high cost producer would choose the contract given by equation (10.2a) and produce Q_H; sales revenue would cover his production cost and he would make a profit equal to the subsidy of X. If he chose the other contract, he would gain $Y + Z$ in terms of the rise in subsidy but, would lose $Y + Z + A$ on his production, so would not choose the contract given by equation (10.2b) over the contract shown in equation (10.2a). Similar logic implies that the low-cost producer would choose the contract in equation (10.2b), produce output Q_L and make profits of $X + Y + Z$.

The contracts in equation (10.2) achieve the first-best level of W, but W is allocated entirely to the producer, since the subsidy the producer receives must be raised by taxation from the consumers; the surplus, S, which the consumers gain from consumption is taxed away from them and given to the producers. In general, one might imagine that the regulator has distributive prejudices in favour of the consumers, or else that there are costs involved in raising tax revenue. Such considerations would change the specification of the problem and also change the optimal contracts; the interested reader can find further details in the references given in section 10.4.

Although very simple, the above model shows why regulatory authorities try not to rely on cost figures prepared for them by producers, and expend effort to try to find accurate cost estimates of their own. The model may be complicated in various ways. One of the most obvious ways is to allow the regulator to set prices for an initial period and then to reset them at a later date once he has observed costs. It might seem that this would allow the regulator to move to the first-best price level once he had an accurate observation of costs, but in reality costs change over time and it is unlikely that the first-best outcome can be achieved in this way. Furthermore, the monopolist will foresee that the regulator will gain knowledge over time and if costs depend on his actions he will manipulate them to convey the message he wants the regulator to receive. Such interactions can become complicated and we shall not discuss them further here, although, again, the interested reader may pursue some of the references given in Section 10.4.

10.3 Procurement with hidden information and hidden action

In this section we consider the problem of a public authority wishing to carry out a major public project, such as providing a new telecommunications network, by commissioning a private-sector company to produce the project. As in the previous section, the public authority does not know the cost function of the private-sector producer. We simplify the problem of the previous section in one dimension by imagining that the scale of the project is fixed, so we do not need to concern ourselves with quantity produced, but complicate it in another since we now allow the cost function of the producer to be affected by the amount of effort he puts into the project. We specify that this effort is unobservable by the purchasing authority and thus add a hidden action problem to the problem of hidden information. We assume that the authority wishes to minimise the cost, C, of commissioning the project as long as the cost is less than S, where S is the value the authority attaches to the project.

The private firm being commissioned to carry out the project has a cost function of the form:

$$C = R - e + f(e) \tag{10.3}$$

where R represents the difficulty of carrying out the project and e represents the effort level of the managers of the firm. The second term on the right-hand side of equation (10.3) shows that costs are reduced directly on a one-for-one basis as the effort level, e, is increased. The final term in equation (10.3) represents the idea that exerting effort yields disutility to the managers of the firm, for which they require the monetary compensation of $f(e)$, which is added accordingly to the cost of the project. We assume that e cannot be negative and that disutility increases at an increasing rate with effort, such that $f' > 0$, $f'' > 0$. Furthermore, we assume that $f(0) = 0$, and that $f(e)$ approaches infinity as e approaches R.[4]

The hidden information problem is that the public authority does not know the value of R, which is the private information of the firm. The hidden action problem is that it cannot observe the effort level, e. As a benchmark, first consider the full-information case, where the authority knows R and can monitor e freely. In this case, the authority wishes to minimise the payment, P, it makes to the firm for carrying out the project, subject to the firm's participation constraint that the payment by the authority to the firm, P, at least covers the cost, C. This latter constraint is derived from the utility function of the firm and is given by:

$$U = P - C \geq 0 \tag{10.4}$$

where U is the utility of the firm.

Setting P equal to C (since the authority does not wish to pay any more than necessary to the firm) yields the following minimisation problem for the authority:

$$\min P = R - e + f(e) \tag{10.5}$$

with respect to e.

The first-order condition for problem (10.5) yields:

$$f'(e^*) = 1 \qquad (10.6)$$

which may be solved for the optimal value of e equal to e^*. Equation (10.6) shows that the optimal value for e is independent of the value of R. The interpretation is easy: e should be increased until the marginal cost of doing so, $f'(e)$, equals the marginal reduction in costs which effort yields – that is, 1. Having solved for e^*, the authority can calculate P^* from equation (10.5) and make the firm the take-it-or-leave-it offer of P^* for carrying out the project.[5] The firm will simply break even by accepting the offer and will optimally choose e equal to e^* for its own selfish optimising reasons. Thus knowledge of R and of $f(e)$ allows the authority to offer the firm a fixed-price contract without the need to monitor e.

Another way of producing the optimal outcome would be for the authority to offer the firm a *'cost-plus contract'*. Under this contract, the authority would offer to pay the firm the net-of-effort cost of producing the project, $N = R - e$, plus a transfer of t equal to $f(e^*)$. If the firm exerts less than e^* effort it must make a penalty payment larger than $f(e^*)$ to the regulator. Given this penalty, the firm will, having accepted the contract, choose the effort level of e^*. The total cost to the authority will therefore be $R - e^* + f(e^*)$ or P^*, as above.

Now introduce the asymmetric information problems of hidden action and hidden information. Assume that the purchasing authority knows that the firm may face one of two different values of R – either R_H with probability p, or R_L with probability $(1 - p)$, where R_H is greater than R_L; this introduces the hidden information problem. The hidden action problem concerns the unobservability of e; the authority is able to observe N, which equals $R - e$, but is not able to observe the component parts, R and e, of N; we assume that the function $f(e)$ is common across the two firms and is known to the government.[6]

Imagine initially that the authority is restricted to offering fixed-price contracts. It will then either offer the single fixed price of P_L or P_H, where these two prices represent, respectively, the fixed prices necessary to allow an efficient firm, with R_L, or an inefficient firm, with R_H, to break even if they set e equal to e^*.[7] We assume that S is greater than P_H (and, therefore, than P_L) so that the authority would purchase the project from either type of firm in a full information world. Clearly, any price less than P_L will be acceptable to neither type of firm, and any price equal to or greater than P_H will be acceptable to both. Under the asymmetric information problem, the authority will not, therefore, have any incentive to offer any price less than P_L or greater than P_H. Nor does it have any incentive to offer a price between P_L and P_H, since such a price simply gives excess returns to the efficient type-L firm without encouraging the inefficient type-H firm to accept the contract.

If the authority sets the price P_L, then only a type-L firm will accept the contract; the authority's expected utility will therefore be $(1-p)(S-P_L)$. If it sets the higher price its expected utility will be $S - P_H$, since both types of firm will accept the contract in this case.[8] The authority will, therefore, set the price P_H only if $S - P_H$ is greater than $(1-p)(S-P_L)$ or S greater than $[P_H - (1-p)P_L]/p$. Assume that this latter condition for the size of S is satisfied, so that the authority would, if restricted to offering a fixed price contract, offer the price of P_H and purchase the project regardless of which type of firm it faced. It is easy to show that in this case the authority can do better than offer the fixed price contract of P_H if it is allowed to offer a cost-plus contract.

Given the information problems facing the authority, the cost-plus contract will now take the form of a transfer of $t(N)$ and the payment of the observed value of N. The transfer $t(N)$ may be determined to be a function of N in such a way as to enable the authority to reduce the expected costs of acquiring the project below the price of P_H that it would cost to ensure production under the fixed-price contract arrangement. The arguments are illustrated using Figure 10.3.

Figure 10.3 plots transfers, $t(N)$, on the vertical axis and net-of-effort costs, N, along the horizontal axis. The straight lines drawn at 45° to the horizontal axis represent the iso-cost lines of the purchasing authority, where the total cost, P, equals $t(N)$ plus N under cost-plus contracts.

The lines U_L and U_H represent, respectively, the break-even indifference curves of the low-cost and high-cost producers. Under full information the purchasing authority would offer whichever type of producer it faced the contract along the producer's break-even locus which would minimise the authority's purchasing cost; that is, it would offer the contract where the indifference curve

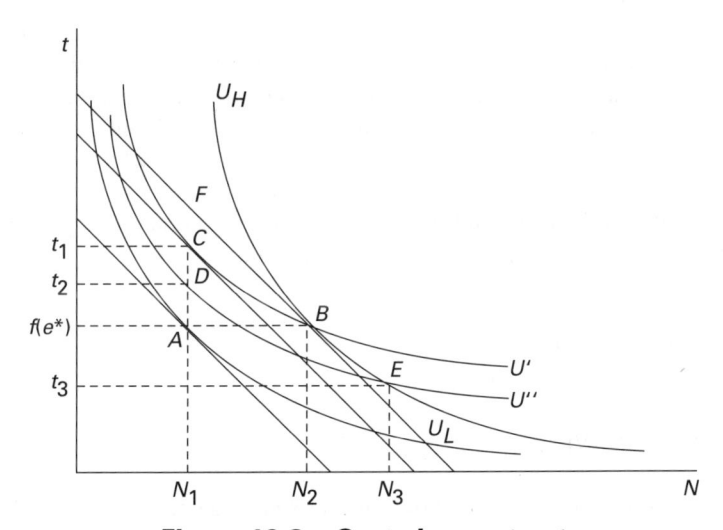

Figure 10.3 Cost-plus contracts

touched the appropriate break-even iso-cost line tangentially. Thus the purchasing authority would offer contract A if facing the low-cost producer, or contract B if facing the high-cost producer at the respective total costs of P_L and P_H. The contracts would cover the specified net-of-effort costs of each type of producer, N_1 or N_2 as the case may be, and offer each the same transfer of t equal to $f(e^*)$; where N_1 equals R_L minus e^* and N_2 equals R_H minus e^*.

Under full information, the producer, whether a high- or a low-cost type, would be induced to exert the optimum effort level of e^* as discussed above with reference to equation (10.6).

Under asymmetric information about costs and effort, contracts A and B will not satisfy incentive compatibility, since the low-cost producer would prefer contract B to contract A. Thus both types of producer would choose contract B and the total cost to the authority of purchasing the project would be exactly as expensive as under the fixed-price contract case discussed above, that is, P_H.

The low-cost producer would prefer the fixed-price case, since at B the level of effort he is expending would be less than the optimal level of e^* and he could improve his own utility at no extra cost to the purchasing authority if offered a fixed-price contract of P_H rather than contract B. Under the fixed-price contract the producer would maximise utility by setting effort equal to e^*, as may be seen easily by maximising U with respect to e under a fixed price.

As an alternative to offering the fixed-price contract P_H to both types, it would be possible for the purchasing authority to offer the cost-plus contracts at B and F under our asymmetric information case; where F is on the same iso-cost as B but vertically above N_1. The low-cost producer would choose F and the high-cost producer, B. The total cost under either contract would be P_H but the low-cost producer prefers F to B and earns higher utility by increasing effort as he moves to F and receives more than full compensation for this extra effort as the transfer, t, is higher at F. This result follows from the slope at B of the low-cost producer's indifference curve, U', being less than the slope of the high-cost producer's indifference curve, U_H, and hence less than the slope of the total cost line (which is tangential to the high-cost producer's indifference curve at B). Better still, however, the purchasing authority can use this information about relative slopes to design a pair of separating contracts which reduces the expected cost of acquiring the project, as we shall now show.

The relative slopes of the indifference curves through point B may be seen by substituting for P and C in constraint (10.4), as follows:

$$\begin{aligned} U &= P - C \\ &= P - R - e - f(e) \\ &= t + N - R - e - f(e) \\ &= t + N - N - f(R - N) \\ &= t - f(R - N) \end{aligned} \qquad (10.7)$$

Total differentiation of equation (10.7) for a fixed value of U yields the slope of the indifference curve as:

$$dt/dN = -f'(R - N) \tag{10.8}$$

Since f' is positive and $R_H > R_L$ then equation (10.8) shows that for a common value of N the slope of the indifference curve of the high-cost producer is steeper than that for the low-cost producer.

Given the above result, it is easy to see that the authority could offer a contract along U' in Figure 10.3 to the north-west of point B which would satisfy incentive compatibility and produce a pair of separating contracts if offered as well as contract B. The best such new contract to offer would be that shown at point C in Figure 10.3, where the indifference curve is tangential to a total cost line. We saw above that at such tangency points the firm chooses the optimal effort level of e^*, so it follows that point C must be vertically above N_1; contract C therefore differs from contract A only by offering a higher transfer to the producer in return for producing net-of-effort costs of N_1.

We may use the assumption of epsilon altruism to argue that a low-cost producer offered contracts B or C would choose the latter. He would make this choice because, although he is indifferent between the two contracts on selfish grounds, contract C reduces the costs and increases the welfare of the purchasing authority.

Thus contracts B and C produce an incentive compatible solution. The outcome is clearly not as good as that under full information, since contract C is more expensive than contract A for the purchaser faced with a low-cost producer, but it is better than offering only contract B or the fixed price of P_H, since contract C is on a lower iso-cost than the iso-cost through point B. However, it is possible to reduce expected costs further and to improve upon contracts B and C by offering contracts D and E.

The benefits to the purchaser of offering contracts D and E may be explained as follows. Offering contract D rather than C to the low-cost producer, reduces the total costs for a purchaser facing a low-cost producer but in order to preserve incentive compatibility it is necessary at the same time to replace contract B by contract E. Point E is located at the intersection of the low-cost producer's indifference curve through D and the high-cost producer's break-even indifference curve and is the cheapest contract to preserve incentive compatibility when offered with contract D while also satisfying the participation constraint of the high-cost producer. Replacing contracts B and C with contracts D and E reduces total costs if the authority faces a low-cost producer and increases them if it faces a high-cost producer. Contracts D and E will be chosen optimally to reduce expected total costs according to the probability, p, which the authority places on the producer being of the high-cost type. Clearly, the more likely is the purchaser to face a high-cost producer, the less desirable it is, *ceteris paribus*, to increase the costs of dealing with such a producer, and so the nearer to B and C will be E and D.

Notice that for any contract such as E, designed for the high-cost producer, an optimal contract D for the low-cost producer must be at a point of tangency between the low-cost producer's indifference curve through D and an iso-cost line; thus the low-cost producer exerts the efficient level of effort of e^*. The high-cost producer is not at such a tangency point and is exerting less than the efficient level of effort; since N_3 exceeds N_2 and we know that N falls as e rises.

Optimally, choosing D and E involves trading off the savings when the producer is a low-cost type against the increase in costs when the producer is a high-cost type: this involves reducing the surplus earned by the low-cost type, who is not forced on to his break-even indifference curve, while reducing the effort level of the high-cost type below the efficient level. The less likely it is that the purchaser faces a high-cost producer, then the further E will be away from B, and the further will the high-cost producer's effort level be below the optimal level of e^*. On the other hand, the less likely it is that the purchaser faces a high-cost producer, then the nearer will D be to A, and the lower will be the rent, or increase in utility above zero, of the low-cost producer.

Since at D the low-cost producer is above his break-even indifference curve, then his participation constraint is not binding, that is his utility level exceeds zero, but his incentive compatibility constraint is binding, that is, he is selfishly indifferent between D and E. The high-cost producer, on the other hand, strictly prefers E to D, so his incentive compatibility constraint is not binding, although since he is on his break-even indifference curve, his participation constraint is binding. This suggests that, formally, contracts D and E may be derived by solving the following constrained-optimisation problem for the purchasing authority:

$$\min \quad E(P) = p(t_3 + N_3) + (1 - p)(t_2 + N_1) \tag{10.9}$$

with respect to: t_3, N_3, t_2 and N_1,
subject to:

$$t_3 - f(R_L - N_1) = t_3 - f(R_L - N_3) \tag{10.10}$$

and

$$t_3 - f(R_H - N_3) = 0 \tag{10.11}$$

where equation (10.10) is the incentive compatibility constraint of the low-cost producer and equation (10.11) is the participation constraint of the high-cost producer.

It is easy to see that the solution to the above problem will also satisfy the participation constraint of the low-cost producer and the incentive compatibility constraint of the high-cost producer. Dealing first with the participation constraint of the low-cost type, notice that this is guaranteed to be satisfied by equation (10.11). If the high-cost producer can break even (as guaranteed by equation (10.11)) then the low-cost type can always make a positive profit just by mimicking the high-cost type and accepting the contract designed for the high-cost type, but incurring lower effort costs than the high-cost type.

Now turning to the incentive compatibility constraint of the high-cost type, which may be written as:

$$t_3 - f(R_H - N_3) \geq t_1 - f(R_H - N_1) \tag{10.12}$$

or

$$t_3 + f(R_H - N_1) \geq t_1 + f(R_H - N_3) \tag{10.13}$$

Looking at constraint (10.13), N_3 is greater than N_1, so that the effort necessary to produce net-of-effort costs of N_1 is greater than the effort necessary to produce N_3. Hence $f(R_H - N_1)$ is greater than $f(R_H - N_3)$ and t_3 is greater than t_1, thus guaranteeing that constraint (10.13) must hold.

We have assumed in setting out the above problem that the value of the project S is greater than $[P_H - (1 - p)P_L]/p$. This assumption was necessary to guarantee that a purchaser offering a fixed-price contract would set the price P_H and always purchase the project whether facing a high-cost or a low-cost producer. This condition is stricter than is necessary for the authority to choose to offer contracts D and E in preference to offering only contract A, or the equivalent fixed-price contract of P_L, which would be accepted only by the low-cost producer. The expected utility of the purchaser offering contracts D and E would be given by $S - p(N_3 + t_3) - (1 - p)(N_1 + t_2)$, which, if both contracts are to be offered rather than offering only contract A must exceed $(1 - p)[S - N_1 - f(e^*)]$. Manipulation of the latter inequality yields:

$$S > (N_3 + t_3) + [t_2 - f(e^*)](1 - p)/p \tag{10.14}$$

Constraint (10.14) will be more likely to be satisfied the higher is the value of S and the higher the value of p – the probability of facing a high-cost producer. The interpretation is straightforward. A high value for S implies that the authority is more likely to want to carry out the project regardless of whether it faces a high- or low-cost producer. A higher value for p implies that the cost savings to be made by offering only contract A will be made with a lower probability and the likelihood of not carrying out the project would increase. Given that t_2 exceeds $f(e^*)$ it follows that S must be greater than $(N_3 + t_3)$ for constraint (10.14) to be satisfied. This is intuitively obvious, since if it were not so the authority would lose whenever it faced a high-cost producer who accepted contract E, in which case it would be clearly better to refuse to offer contracts D and E and to offer only contract A and deal only with the low-cost producer.

Notice that condition (10.14) is less strict than our earlier one which guaranteed that a purchaser able to offer only a fixed-price contract would offer P_H in order to guarantee a purchase. In terms of cost-plus contracts the stricter condition is equivalent to saying that the purchaser prefers offering only contract B to offering only contract A. Since the purchaser prefers offering the separating contracts D and E to offering only contract B it must also prefer offering the separating contracts to offering only contract A, so that condition (10.14) must also hold whenever the stricter condition holds.

We have presented the separating contracts as cost-plus contracts where the transfer in excess of costs, t, depends on the level of net-of-effort costs incurred, N. Alternatively, it would be possible to specify a cost payment and a transfer dependent on the announced value of the efficiency parameter, R, made by the firm on taking up the contract. If a transfer of t_2 and a cost payment of N_1 was specified for an announcement of R_L, and t_3 and N_2 were specified for an announcement of R_H, the low-cost producer would choose to announce R_L and locate at point D in figure 10.3, while the high-cost producer would announce R_H and locate at point E. Thus, in practice, offering cost-plus contracts or asking producers to reveal their efficiency parameters can have the same outcomes as long as the contracts are designed appropriately and are honoured by the purchaser and producer. The purchaser will, however, face the temptation to renege on the contracts offered and must resist this temptation if it is to gain from designing optimal contracts.

The purchaser must not succumb to the temptation when observing costs of N_1 to renege on its contract and reduce the transfer from t_3 to $f(e^*)$ in order to reduce the total cost of the project by cutting the supernormal profits, or rents, of the low-cost producer. This follows since if the low-cost producer realised that the purchaser was going to renege in this way he would choose to locate at E rather than at D and the purchaser would be made worse off than if he had not offered incentive compatible separating contracts in the first place.[9]

Under the information-revealing type of contract the need to abide by the contract is a little less obvious. Here the problem is that a high-cost producer revealing R_H would gladly renegotiate its contract to move from E towards B as long as the purchaser offered some of the resultant cost savings (resulting from the improved efficiency in effort level) to the producer to increase his utility above the break-even level.[10] The trouble with such renegotiation is that it destroys incentive compatibility. This occurs because a low-cost producer, knowing that such renegotiation takes place, no longer chooses between D and E when announcing his efficiency parameter, but between D and the anticipated renegotiation point for the high-cost producer. The low-cost producer will prefer the anticipated renegotiation point to either D or E and would be given an incentive to mimic the high-cost producer if he thought the incentive contracts would be renegotiated once accepted. The purchasing authority must therefore gain a reputation for not being prepared to renegotiate if it is to reap the gains from incentive contracts. Thus, unlike in the story of the wisdom of Solomon, with which we introduced the idea of asymmetric information, it is usually not wise for an authority to break its word or renegotiate contracts.

Finally, notice that we have examined contract-based responses to the purchasing problem under asymmetric information. Another way of tackling the problem would be for the purchaser to ask firms to place tenders for the project and to devise the method of tender to try to achieve an optimum outcome. This method is important and deserves coverage, but since the technical analysis of this case is similar to that of auctions, with which we deal in the next chapter, we shall not deal with it here.[11]

10.4 Recommended reading

Regulation is a vast topic and we have only touched upon it in this chapter. For an accessible survey of the literature see Vickers and Yarrow (1988, Ch. 4). For a comprehensive, detailed and advanced coverage see the important book by Laffont and Tirole (1993). Seminal papers in the area include Baron and Myerson (1982), which modelled regulation under hidden information as in section 10.2 above, and Laffont and Tirole (1986), which examined a problem with hidden action and hidden information as in Section 10.3 above. Helm (1994) offers an interesting account of theory and practice in relation to British utility regulation.

10.5 Problem

Problem 10.1

A public authority wishes to commission a public project which it values at $S = 500$. The authority wishes to minimise the cost, C, of the project. The private firm being commissioned to carry out the project has a cost function of the form:

$$C = 300 - e + (e^2)/20 \qquad (10.15)$$

where e represents the effort level of the managers of the firm.

(a) If the authority knows the cost function (equation (10.15)) and can observe effort, calculate the optimal cost-plus contract (t^*, N^*) it can offer the firm, subject to the firm's participation (break-even) constraint being satisfied; where $N = 300 - e$ and $t = (e^2)/20$. What is the optimal cost of the project?

(b) If the cost function (equation (10.15)) were replaced by

$$C = 304 - e + (e^2)/20 \qquad (10.16)$$

how would the optimal values for t and N change?

(c) Now consider the case where the purchaser is unable to observe effort and is unsure of the cost function of the producer. The purchaser attaches equal probabilities to the cost function taking the form of equations (10.15) or (10.16). Calculate the optimal separating contracts (t_L, N_L) and (t_H, N_H), designed for the low-cost and high-cost producers respectively. Compare the optimal values for (t_L, N_L) and (t_H, N_H) for each type of producer under separating contracts with the optimal values that would be set by a purchasing authority in a full-information world. Say what the implications are of the asymmetric information problem for effort levels and firm rents.

Hint: look at equations (10.9) to (10.11) before tackling this part of the question.

(d) Verify that the separating contracts derived in (c) above satisfy any relevant incentive compatibility or participation constraints not used in your optimisation procedure.

(e) Is the expected welfare of the purchasing authority facing the asymmetric information problem higher if it offers the pair of separating contracts determined in (c) above or if it offers only the contract determined in part (a) above?

CHAPTER 11

Auctions

11.1 Overview

An *auction* is a form of market with a well-defined set of rules specifying how the participants in the market must behave. There are several different types of auction and these are widely used for selling such diverse items as antiques, works of art, motor vehicles, government debt, fish, meat, cattle, diamonds, houses, land and the rights to exploit natural resources such as oil. There are also auctions for the rights to deliver goods or services – for example, a government may ask for construction companies to bid for the rights to construct a new stretch of highway. These latter auctions are known as *buyers' auctions,* since the government will buy from the winning bidder rather than sell to the winning bidder as in the previously described *seller's auctions.* For simplicity, the general discussion that follows will be phrased in terms appropriate for sellers' auctions, but a similar analysis may be applied to buyers' auctions.

Given that auctions are widely used it is natural to inquire why and to seek to examine the different types of auction rules to see if they affect the outcome of the auction. We shall see that auctions are primarily a response to a hidden information problem, since a seller of something may not know the potential buyers, whom we shall also call *bidders,* for reasons that will soon become clear, or their valuations of the item for sale. They are also partly a response to hidden actions; for example, the hidden actions might involve collusive behaviour by potential buyers which is designed to reduce the price paid to the seller. We shall also see, perhaps surprisingly, that apparently very different auction mechanisms can, under certain circumstances, produce the same expected revenue for the seller and so satisfy the *revenue equivalence theorem.*

Section 11.2 covers definitional aspects by introducing the possible information asymmetries present when an item is to be offered for sale and also introducing the main types of auctions to be discussed. Section 11.3 makes use of an extended example to analyse the behaviour of the participants in the different types of auction and to present the revenue equivalence theorem. Section 11.4 presents another simple example to introduce the idea of an *optimal auction*;

that is, an auction designed to maximise the expected revenue of the seller. Section 11.5 introduces the *winner's curse*, or the idea that in some circumstances a bidder, having succeeded in winning the auction, might find that he has 'paid too much' for the item he has just bought. Section 11.6 offers some recommended reading and section 11.7 presents problems for the reader to solve.

11.2 Auctions and information problems

When discussing auctions it is necessary to be careful about both the information structure assumed in the problem and about the specific rules of the type of auction being considered.

The information structure is usually considered to be one or other of two extremes known as *private-value* or *common-value* cases. In the private-value case, the assumption is that each potential buyer of some item knows how much he would be willing to pay for the item; that is, he knows the *value* of the item to him. The hidden information problem in this case occurs because the value of the item differs across potential buyers and the value placed on the item by any individual buyer is his own private information and is not known to either the seller or any other potential buyer. It is usually assumed that the value held by any potential buyer may be viewed by the seller and other potential buyers as a random variable drawn from some probability distribution. Assuming that the value for each potential buyer is given by an independent drawing from a common probability distribution yields the *symmetric independent private values* or *SIPV* model which has been widely discussed in the literature on auctions.

In the common-value case, the value of the item for sale is common to each of the bidders but no bidder is certain of this value and, instead, possesses only an estimate of it; we call this estimate his *valuation*. Bidders therefore face a problem of uncertainty and there is also a hidden information problem since the valuation of each bidder is his own private information and reveals nothing of the valuations of other bidders.

Auctions in the real world probably display a more general information structure known as the *correlated-value* case. In this case the values of different bidders are correlated. Alternatively, it could be the case that the unknown value is common and it is the bidders' estimates of this value – that is, their valuations – that are correlated; for example, a bidder for drilling rights on some oil field who acquires favourable information from private testing is likely to infer that other bidders have also received similar information.

We shall concentrate on the private-value and common-value cases in what follows. These cases are easier to analyse than the correlated-value case and seem to be reasonable simplifications. They do not omit too much that is interesting and they are capable of helping us to understand a complicated world more easily. For example, consider an auction of a house. We would model this as a private-value auction, since the value to any bidder may be judged to be primarily given by factors specific to him, such as the proximity to his place of work or to his mother-in-law's house, while we would model the auction of rights to drill for oil as a common value auction, since the value of the rights may reasonably be assumed to be common across different companies in the bidding and we may ignore the complications of allowing for correlated valuations. We might note here that the common value in a common-value case may either be known or unknown to the seller; for example, a government selling drilling rights might not know the value of those rights any more than any bidder but, on the other hand, if I filled a jar with coins and allowed my students to bid in an auction for the jar after seeing it for only 30 seconds, then it would be reasonable to assume that I knew the value of the coins in the jar but that my students each had only an estimate of its true common worth.

We shall consider four different types of auction: *English*; *second-price sealed-bid*; *Dutch*; and *first-price sealed-bid*. In the English auction, sometimes also known as the *first-price open-cry* auction, the auctioneer invites open or oral bids which can be observed by all participants. A bid must exceed the previous bid and bidding continues until no one is willing to go above the last bid made. The bidder who made the final bid gets the object being sold, at the final bid price. This is probably the most common type of auction to be observed in the real world and is what many people would imagine an auction to be.

In the *second-price sealed-bid* auction the auctioneer asks for bids to be made, in ignorance of the bids of other bidders, on the understanding that the object will be sold to the highest bidder (usually by some set closing date) at a price equal to the bid made by the second highest bidder. This type of auction, unlike the other three, is not commonly observed in practice. It was devised by Vickrey (1961), in an important study of auctions, as a sealed-bid auction capable of producing behaviour and results equivalent to the open-bid English auction. The second-price sealed-bid auction is, therefore, sometimes known as a *Vickrey* auction.

In the *Dutch auction* the offer price starts at a price believed to be higher than any bidder is willing to pay and is systematically lowered by the auctioneer, or some mechanical device or clock which ticks down, until one of the bidders signals that he is willing to pay the currently indicated price. Hence, in a Dutch auction, the first bid wins. Dutch auctions, although not as common as English auctions, are used, for example, to sell flowers in the Netherlands, fish in Belgian ports or tobacco in Ontario.

In the *first-price sealed-bid* auction the auctioneer asks for bids to be made, in ignorance of the bids of other bidders, on the understanding that the object will be sold to the highest bidder (usually by some set closing date) at a price equal to the bid made by the highest bidder.[1]

In sealed-bid auctions we assume that in the event of a tie, where two or more bidders make the same bid, the winner will be chosen at random. In the English auction a tie is unlikely, since only ascending bids are requested, but should two bidders call out the same price simultaneously we may again assume a random method is used to choose between them (assuming, of course, that neither is prepared to go higher and become an outright winner). In the Dutch auction we might also assume a random choice method if two or more buyers 'stop the clock' simultaneously. Random choice methods may, however, cause a problem in a Dutch auction since a bidder will, as we shall see later, choose to postpone making his bid even though the price has dropped below his reservation value in such an auction and therefore will have an incentive to try to jump in once another bidder stops the clock if the current price is less than his reservation value. Those carrying out such auctions must therefore take pains to ensure that they can distinguish who bid first; for example, by providing bidders with stop buttons and monitoring the time at which the buttons are pressed – hence the object for sale goes to the person registering the first bid, so bidders may need quick reflexes!

Auctions are sometimes also classified according to whether they are *discriminatory* or *non-discriminatory*, or whether they are *refundable* or *non-refundable*. The first criterion refers to whether the winning bidder pays his actual bid, in which case the auction is discriminatory, or an amount determined by the second highest bidder, in which case the auction is non-discriminatory.[2] An English auction may seem to be discriminatory but is better imagined as non-discriminatory, since the winner need pay an amount only slightly higher than the value of the second highest bidder. The refundable/non-refundable distinction refers to whether losing bidders pay some (possibly all) of their bids to the seller or auctioneer; if they do, the auction rules are non-refundable and if they do not, they are refundable.[3] We shall consider only refundable auctions in what follows.

11.3 Private value auctions and the revenue equivalence theorem

Let us consider a specific example in order to concentrate our minds. Ben has inherited a large and (he hopes) valuable house from a long-lost uncle who lived in a distant part of the country. Ben is currently living in a tiny and unpleasant flat. He plans to sell the house and buy another in the area where he is living.

Since Ben and the house are both located in England the usual procedure for selling a house is to approach a local estate agent who will provide an estimate of the market value of the house, thus helping Ben with his first information problem since he is very uncertain of house prices in the distant region where his uncle used to live. The estate agent, of course, advises Ben that he is also unsure of the potential buyers' values, but that the lower the price, the quicker the house is likely to sell. Ben and the estate agent settle on what seems to be a sensible price and the estate agent advertises the house by providing details of it in his office, displaying them in his window and placing advertisements in local newspapers, all actions clearly designed to solve another information problem – that is, potential buyers must be provided with the information that the house is up for sale and given details of its characteristics. The estate agent explains to Ben that the housing market is sluggish and that he should not necessarily expect a quick sale.

Six months later Ben has still not sold his uncle's house, despite cutting the asking price by 10 per cent in an attempt to attract a buyer. He approaches the estate agent and asks if there is any way he can achieve a very quick sale without cutting the price to an extremely low level, which he fears might signal to prospective buyers that the house is really not a very desirable property. The estate agent says that he sometimes holds auctions to sell property and that this method of selling usually produces a sale very quickly. Since all this is happening in England the estate agent offers Ben two ways of auctioning the house, either an English auction to be held on a specific date sufficiently far in the future to allow potential buyers to be advised of the date and place of the auction, and to give them time to inspect the house if they wish, or to ask for sealed bids by a specified future date with the house going to the highest bidder on that date. He also offers Ben the possibility of specifying a *minimum* or *reserve price* which must be bettered if the house is to be sold. Ben thinks that the idea of an auction is a good one but asks for a few days to think over which type to use and what value, if any, to set as a reserve price.

Ben, being an inquisitive person, goes to the library and studies auction theory from a textbook to try to find some information to help him decide how to instruct the estate agent. He is pleased to find that one of the main advantages claimed for auctions is that they do indeed offer speed of sale. He finds information on all four types of auction we listed earlier and studies them all even though he has only been offered a choice over two of them by the estate agent. He soon learns, however, that the estate agent has not greatly reduced his choice by offering him only two of the four standard auction types, since the English auction is equivalent to the second-price sealed-bid auction and the Dutch auction is equivalent to the first-price sealed-bid auction. He finds this result to be quite surprising, but is even more surprised to discover the revenue equivalence theorem which he correctly interprets as saying that all auctions that award the item for sale to the highest bidder produce the same expected revenue! He decides to check these results very carefully to make sure that he has properly understood them.

Ben finds that, as usual with theorising, the results produced depend upon the assumptions made. He is happy to grant, for simplicity's sake, that the symmetric independent private value model is a reasonable description of his problem, since the bidders are likely to be ignorant of the values placed on the house by other bidders and all might reasonably assume that each bidder's value was given by an independent drawing from a common distribution.

He quickly convinces himself that the Dutch auction and the first-price sealed-bid auction, which we recall was one of the two types offered by the estate agent, are identical. In the Dutch auction case each bidder, in ignorance of the values of other bidders, has to calculate when to shout 'Mine' as the price falls. Ben sees that the bidder could work out this price before the auction begins and write it on a piece of paper and that, since the bidder learns nothing at the auction other than whether he is the first to call 'Mine' or not, this price would be exactly the price the bidder would bid in a first-price sealed-bid auction. The house will be sold to the bidder who chooses the highest price at which to shout 'Mine', or who places the highest sealed bid. Since the price at which the bidder shouts in the Dutch auction will be equivalent to the bid he would make in the first-price sealed-bid auction, Ben agrees that he could expect to receive the same selling price from either a Dutch or a first-price sealed-bid auction, thus accepting the revenue equivalence theorem when applied to these two auction types.

Next, Ben convinces himself that the English and the second-price sealed-bid auctions are equivalent. He realises that the two types of auction differ in the sense that in the former bidders are able to learn about, and respond to, opposing bids as the auction progresses, which they cannot do in the sealed-bid auction. However, he also sees that in the English auction a bidder's optimal strategy is to continuing upping the bidding (by as small an amount as possible) as long as the highest price bid by an opponent is less than the value he places on the house. Thus in the English auction the house will be sold to the highest bidder at a price just marginally above the value placed on it by the second highest bidder.

In a second-price sealed-bid auction the optimal strategy of any bidder is to bid his true valuation in the knowledge that, unless a tie occurs, he will gain the house at the cost of only the second highest bid. No bidder has any incentive to bid less than his true value since to do so would only lower his chances of winning the auction and would not reduce the price he would pay for the house if he were to win. Nor does any bidder have any incentive to bid above his private value, since this would help him to win the auction only if the second highest bid is also above his private value, so that he ends up paying more for the house than the value he attaches to it. Thus all bidders bid their true private value and the house will be sold to the highest bidder at a price equal to the value placed on it by the second highest bidder; that is, at a price only marginally different from the price reached in an English auction. Ben therefore agrees that the two types of auction produce equivalent outcomes, at least to a close approximation in the real world, and accepts the revenue equivalence theorem when applied to them.

Ben feels that although the Dutch and first-price sealed-bid auctions are equivalent to one another, and the English and second-price sealed-bid auctions are equivalent to one another, he should choose the first-price sealed-bid auction rather than the English auction to maximise his expected revenue from selling the house. His reasoning for this choice goes along the lines that selling at the highest bid price must surely be better for him than selling at the second-highest bid price. 'After all,' he argues to himself, 'if you are going to sell at the second highest price bid then why not sell to the second highest bidder?' Then, being a thoughtful and flexible character, he thinks, 'No, that can't be right. If you sell to the second highest bidder the rules of the auction have changed and it might mean that it is no longer optimal to bid your true value. The optimal bid clearly depends on the rules of the auction so I had better check what would be the optimal bid in a first-price sealed-bid auction to see what bids I can expect. I don't suppose the optimal bid will be to bid your full private value, since although this maximises your probability of winning it means you really gain nothing since you pay a price equal to your full private value of the house. Surely the optimal bid will be somewhere below your private value so that if you win then you make a gain from doing so. On the other hand, you must also take into account that as you reduce the bid you also reduce your chances of winning. I wonder how you work out this trade-off between bid price and probability of winning?' Ben was, of course, quite correct to start thinking along these lines and he began to read the textbook more carefully to see why it argued that all four auction types would produce the same expected revenue.

Ben found it easier to work with a specific case of the symmetric independent private values model where the individual private values are drawn from a uniform probability distribution on the interval [0, 1] and there are a given number, n, of bidders. At first he thought this was a silly model since the value of his house did not lie between 0 and 1, but after a while he realised that he did have an idea about the top and bottom values that bidders might attach to the house and that by a suitable choice of scale and origin these values could be normalised to lie between 0 and 1 for simplicity.

Ben read that it was easy to calculate the optimal bid of the ith bidder in a first-price sealed-bid auction; which is, of course, equivalent to the price at which he would shout 'Mine' in a Dutch auction. The argument, which Ben did not find all that easy to follow despite the claim of the textbook writer, went something along the following lines.

Let the ith bidder have a private value of the house equal to v_i. Assuming each bidder, for whom bidder i may be treated as the representative, to be risk-neutral, he wishes to maximise his expected return from the auction, which may be denoted H_i. Thus, bidder i wishes to choose his bid, b_i, to maximise the difference between the value, v_i, which he places on the house and his bid, multiplied by the probability of the bid being the winning bid (since, if it is not, he neither gains nor loses anything); that is:

$$H_i = (v_i - b_i)p(b_i) \qquad (11.1)$$

where $p(b_i)$ is the probability of b_i being the winning bid.

Ben accepted, without formal proof since it seemed intuitive to him, that bids would depend on the values placed on the house by the bidders in such a way that the higher the value held by a bidder, the higher the bid he would be prepared to make. Hence we may write that

$$b_i = f(v_i) \qquad (11.2)$$

where f represents some relationship, to be determined soon, between b_i and v_i which satisfies $f'(v_i) > 0$.

Since each potential bidder faces the same type of maximisation problem (apart from each, of course, having a different value of v), we posit that each will have the same relationship, $b = f(v)$, between b and v. Therefore, we may write the probability of b_i being the winning bid as follows:

$$p(b_i) = \text{Probability}[f(v_{n-1}) < b_i] \qquad (11.3)$$

where b_{n-1} equals $f(v_{n-1})$ is the highest bid from the $n - 1$ bidders excluding bidder i, and v_{n-1} is the highest value held by any of the $n - 1$ bidders excluding bidder i.[4] Equation (11.3) may be rewritten as follows:

$$p(b_i) = \text{Probability}[v_{n-1} < g(b_i)] = C[g(b_i)]^{n-1} = C[v_i]^{n-1} \qquad (11.4)$$

where g is just the inverse of f, so that $g(b_i)$ represents the value of v_i that leads to bidding the amount b_i and, given the assumption of a uniform distribution, $C[v] = v$, for $v \in [0, 1]$ is the probability of drawing a value of less than v from the uniform distribution.

Equation (11.4) has the interpretation that, given that all bidders choose their bids according to the relationship b equals $f(v)$, the probability of b_i being the winning bid is equal to the probability that v_i is the highest value held by any of the n bidders. For example, with n equal to 2 and v_i equal to 0.5, there is a 0.5 probability that the v of the remaining bidder is less than 0.5; with 3 bidders, the probability that 0.5 is the highest value becomes the probability that both the remaining bidders have v's below 0.5, which produces $C[v_i]^{n-1}$ equals $(0.5)^2$ or 0.25 and so on for other values of n. Of course, to make these calculations, each bidder must know the number of bidders involved in the auction. Ben thinks that with sealed bids entered by individuals it is unlikely that this assumption will hold, but he is willing to grant it for the sake of simplicity.

Armed with equation (11.4), it is possible to solve explicitly for the optimal bids. This can be done by differentiating equation (11.1) with respect to b_i to find the following first-order condition:

$$p'(b_i)(v_i - b_i) - p(b_i) = 0 \qquad (11.5)$$

Using equation (11.4) to evaluate $p'(b_i)$ by differentiation using the chain rule as $(n - 1)C[g(b_i)]^{n-2}c[g(b_i)]g'(b_i)$, where c equals C', and substituting this result and $p(b_i)$ equals $C[g(b_i)]^{n-1}$ in equation (11.5) yields:

$$(n - 1)C[g(b_i)]^{n-2}c[g(b_i)]g'(b_i)(v_i - b_i) - C[g(b_i)]^{n-1} = 0 \qquad (11.6)$$

Dividing through equation (11.6) by $C[g(b_i)]^{n-2}$ yields:

$$(n - 1)c[g(b_i)]g'(b_i)(v_i - b_i) - C[g(b_i)] = 0 \qquad (11.7)$$

Noting that for the uniform distribution $\in [0, 1]$ we have $C[g(b_i)]$ equals $g(b_i)$ and $c[g(b_i)]$ equals 1, as well as noting that v_i may be replaced by $g(b_i)$ by definition, then equation (11.7) may be simplified to yield:

$$(n-1)g'(b_i)(g(b_i) - b_i) - g(b_i) = 0 \tag{11.8}$$

The inverse of the relationship v_i equals $g(b_i)$, which satisfies equation (11.8), will determine the optimal relationship between b_i and v_i, but what is it? Simply try the linear relationship v_i equals gb_i to see if such a relationship satisfies equation (11.8) and it is possible to see that it will do so for a certain value of g.[5] In other words, replace $g(b_i)$ by gb_i and $g'(b_i)$ by g in equation (11.8) to produce equation (11.9), which may be manipulated to yield equation (11.10):

$$(n-1)g(gb_i - b_i) - gb_i = 0 \tag{11.9}$$

thus

$$g = 1 + 1/(n-1) = n/(n-1) \tag{11.10}$$

Hence v_i equals gb_i equals $[n/(n-1)]b_i$ satisfies equation (11.8) and inversion yields the optimal bid b_i as a function of v_i as

$$b_i = [(n-1)/n]v_i \tag{11.11}$$

Equation (11.11) says that for a first-price sealed-bid auction, the optimal bid b_i in the SIPV model with bidders' values drawn from the uniform distribution over [0, 1] is simply calculated as $[(n-1)/n]v_i$. With two bidders, each should bid half of the value he attaches to the house; with three, each should bid two-thirds of his value, and so on. Equivalently in a Dutch auction, the ith bidder should plan to stop the clock when the price reaches $[(n-1)/n]v_i$.

Notice that as the number of bidders increases, the optimal bids move closer to the values held by the bidders, and also that the bidder with the highest value will make the highest bid and be the winner. It therefore follows that the outcome of the auction will be Pareto optimal, since the object for sale will go to the person who values it the most; this result is not always the case in Dutch or first-price sealed-bid auctions and depends on our specific example.

In order to illustrate the expected revenue theorem we need to know the expected revenue produced by Dutch or first-price sealed-bid auctions when bidders adopt the optimal bidding strategy. The answer is clearly $[(n-1)/n]$ times the expected value of the highest value held by a bidder. Given our assumption of a uniform distribution over the range [0, 1] the expected value of the highest value is $[n/(n+1)]$, so the expected revenue is $[(n-1)/(n+1)]$.[6] We now need to compare this with the expected revenue produced by an English auction or the equivalent Vickrey auction.

Recall that in the Vickrey auction the optimal bid is simply to bid one's true value and that the bidder with the highest value wins and pays a price equal to the second highest value, while in an English auction the bidding stops when the bidder with the second highest value stops bidding after the price has reached his value and the object for sale goes to the winning bidder at a price approximately equal to the second highest value. In either case, the outcome is always

Pareto optimal, since the object for sale goes to the bidder with the highest value of it. In both cases the expected revenue is equal to the expected value of the second highest value held by the sample of bidders. Given the uniform distribution over [0, 1] it turns out that this expected value is exactly equal to $[(n-1)/(n+1)]$, thus verifying, for our example, the revenue equivalence theorem.[7]

The revenue equivalence theorem does not mean that whichever auction type Ben chooses to hold will produce the same revenue for him. It means that each type will produce the same *expected* revenue, but on any one occasion it is likely that the two different subsets of auctions will produce different proceeds. For example, imagine that there are just two bidders for the house, one with a value of 0 and the other with a value of 1. In an English auction, the low-value bidder would drop out as soon as the bidding started and the house would go for an amount marginally higher than zero to the high value bidder, while in a Dutch or first-price sealed-bid auction, the high-value bidder would win, but after bidding 0.5 and paying 0.5. On the other hand, if the low value bidder held a value of 0.98, the reader ought to be able to calculate that in a Dutch auction the revenue would remain fixed at 0.5, but the higher revenue would be produced by the English or Vickrey auction. It seems from these two examples that the actual revenue might be more variable in an English or Vickrey auction than in the Dutch or first-price sealed-bid cases and this is indeed the case. Thus a risk-averse seller may prefer the Dutch or first-price sealed-bid auctions. Ben, however, is willing to accept risk and feels indifferent between the two auction types offered by the estate agent. He thinks about tossing a coin to determine his choice but decides there must be a better way than that and so reads on a bit further to see if he can find any reason for choosing one type of auction over another.

One argument that Ben thinks may be relevant to him is the question of whether bidders can collude to reduce competition among themselves and reduce the price he will receive. Such collusion is termed forming an *auction ring*. In the real world, with large amounts of money at stake, bidders do have an inducement to form such rings, which usually work by the ring members agreeing together on who should win the auction and the other members refraining from placing high bids so that the winner pays less than he would otherwise have done. In return for this restraint the ring members receive a payment from the winner which is their share of the saving he has calculated he has been able to make as a result of the activities of the ring. Such rings obviously work against the interest of the seller and in favour of the bidders.

A reader with a little imagination will be able to guess that ring members may have an incentive to claim a falsely high value for the object being sold so as to claim a high side payment from the winner and that rings face problems in enforcing agreements reached by the members.[8] Auctioneers sometimes try to maximise such enforcement problems by keeping the identity of the winner of the auction a secret (although, of course, this is difficult in open-cry auctions). These enforcement problems are greater in Dutch or first-price sealed-bid auctions than in English or Vickrey auctions. The reason is that, in the former two types of auction the winner designated by the ring puts in a low bid or waits a long time before shouting 'Mine', and those designated to put in lower bids or refrain from stopping the clock are tempted to renege on the ring agreement by putting in a bid just slightly higher than the designated winner's bid or shouting 'Mine' just before the designated winner does. In an English or second-price sealed-bid auction the designated winner can bid his own true valuation and win the auction even if other ring members cheat and put in higher bids than the ring agreed; hence those designated to place low bids or refrain from bidding by the ring cannot gain by reneging, since, unless they have not revealed their true value of the object for sale to the ring, they will still be beaten by the designated winner's bid, and so the ring agreement is self-enforcing.

Ben finds this reason to favour the first-price sealed-bid auction over the English auction to be rather unconvincing, for two reasons. First, he thinks that if a ring does exist it will probably be prepared to enforce its agreements by meting out severe (even illegal) punishments to those who renege. Second, he is more concerned about the honesty of the estate agent in carrying out the first-price sealed-bid auction.

The reason for Ben's concern about the estate agent is that he thinks the agent might open the bids and then let a friend or associate know the value of the highest bid from the other bidders. If the friend or associate then wishes he can put in a marginally higher bid than any other bidder and win the auction at a price just marginally higher than the second price. At first sight this might seem just to convert the first-price sealed-bid auction into a second-price sealed-bid auction and to preserve revenue equivalence. Ben, however, sees that this is not so, since the bidding behaviour in the two auction types is different and the second highest bid in a first-price sealed-bid auction will be lower than the second highest bid in a second-price sealed-bid auction as long as the bidder with the second highest value of the house follows his optimum bidding strategies in each type of auction. Thus, collusion between the estate agent and a bidder could reduce seriously the expected revenue for Ben in a first-price sealed-bid auction. In an English auction, he reasons, if the estate agent's friend wants to win the auction he will have to make the highest bid and will genuinely have to have the highest value of the house and cannot be helped by the estate agent. This problem of collusion between auctioneer and bidder is particular to the first-price sealed-bid auction and would not appear in a Dutch auction, where the auctioneer would be unable to gain knowledge of bidders' values to assist his associate. Ben's choice is between English and first-price sealed-bid auctions, so he chooses the English auction for the sale of his house.

Before closing the textbook Ben noticed a section on optimal auctions which claimed that it is possible to increase expected revenue by designing an optimal auction. He skipped over this section and soon decided that he would not try to design one for the sale of his house. He made this decision not just because he had been offered a choice of only two auction types by the estate agent, but also because he thought the section looked too complicated. Even if he could, in principle, design an optimal auction he was not sure that potential bidders would understand the rules properly and might even be dissuaded from bidding as a result. Indeed, as Ben realised, a strong argument in favour of the English auction is that anyone ought to be able to understand the rules and work out for themselves that bidding up to one's value is the best thing to do.

Ben saw an argument that expected revenue could be increased by using a reserve price and that setting such a reserve price violates a necessary assumption for the revenue equivalence theorem that the object for sale is always awarded to the highest bidder. He decided not to pursue that strand of analysis or to try to calculate a reserve price designed to maximise his expected revenue, since the higher the reserve price the less likely it would be that the auction would lead to a sale of the house. 'No,' he thought to himself, 'I want a sale of this house as long as the price rises above £X,000. If this price is not reached I might as well keep the house and let it to a tenant. I shall simply set £X,000 as my reserve price in an English auction and be done with it.'

Relaxing the assumptions

Before moving on to consider optimal auctions it is worthwhile emphasising here that the revenue equivalence theorem depends on the assumptions of the symmetric independent private values model. When these assumptions are relaxed the equivalence theorem breaks down. We have already seen that when there is collusion among bidders there may be a reason for the seller to prefer a Dutch or first-price sealed-bid auction if the ring faces an enforcement problem, or that if there is collusion between the auctioneer and a bidder the seller may prefer an English auction to a first-price sealed-bid auction. Thus the revenue equivalence theorem is not robust in the face of collusive behaviour.

A key assumption for the revenue equivalence theorem is that bidders are risk neutral. When bidders are risk averse the seller gains a higher expected revenue from holding a Dutch or first-price sealed-bid auction rather than an English or Vickrey auction. The reason is simple. In the English or Vickrey auction, risk aversion affects neither the optimal bidding strategy nor the expected revenue. In a Dutch or first-price sealed-bid auction, however, the optimal strategy involves *shading* one's bid – that is, reducing it below one's value – and trading off the probability of winning against the size of the surplus if one wins. Risk averse bidders, therefore, are reluctant to shade their bids and reduce their chances of winning; thus they tend to bid higher than risk neutral bidders would and so increase expected revenue for the seller. Similar arguments follow from recognising that bidders might be both risk averse and uncertain about the number of bidders.

Another key assumption is that bidders' values are independent. Replacing this by the assumption that values are correlated positively gives the seller a reason to prefer an English auction. Also important for the revenue equivalence theorem is the assumption of symmetry. Relaxing this assumption makes it impossible to provide general arguments for ranking auction types since the ranking then depends on the specific heterogeneity introduced when relaxing the symmetry assumption.

Similarly, it is impossible to provide sharp results if we relax the assumption that bidders follow optimal strategies; if bidders find the auction rules or optimal strategies difficult to understand, then their behaviour becomes difficult to model and it may be that the revenue equivalence theorem will no longer continue to hold. Also notice that the theorem applies to single unit auctions and no longer applies if the seller is offering for sale more than one unit of a good or if the quantity to be sold is endogenous and depends on the price. Finally, notice that the theorem is not robust to minor modifications of the auction rules such as setting a participation fee on bidders or setting a minimum reserve price.

11.4 Optimal auctions

Although Ben ruled out designing an optimal auction on the grounds of practicality it is worth considering the idea here. We shall see that designing an optimal auction provides an excellent example of the use of the Revelation Principle we first encountered in Chapter 4.

In order to keep the analysis as simple as possible we shall consider a case of the symmetric independent private values model with only two risk neutral bidders. Each bidder, of course, knows the value he places on the single object for sale, but this value is assumed by the seller and the other bidder to be a random variable which takes either a high value, v_H, with probability $(1 - p)$ or a low value, v_L, with probability p. In order to concentrate our minds even more, let us assume that v_H equals 2, v_L equals 1 and p equals 0.5.

The task of the seller is to design an auction that will maximise his expected revenue. Before carrying out this task let us find a benchmark figure for expected revenue by finding what value would be produced by an English auction with no reserve price. In this case the expected revenue, $E(R)$, would be given by:

$$E(R) = (0.75)1 + (0.25)2 = 1.25 \qquad (11.12)$$

where the first term on the right-hand side of equation (11.12) is the probability (0.75) that the seller faces either two low-value bidders or a low-value and a high-value bidder times the price of 1 which would be achieved in such cases, and the second term is the probability of both bidders being of the high-value type (0.25) times the price of 2 which would be achieved in this case. Given the revenue equivalence theorem, we know that the expected revenue of 1.25 would remain unchanged if the seller held any of the other three standard auction types. Can this figure be bettered by designing an optimal auction?

In designing an optimal auction we can make use of the revelation principle, which says that any outcome achieved by a mechanism which gives one or more of the bidders an incentive to lie, can be achieved also by a mechanism which provides no such incentive. Hence we need only examine mechanisms which induce truth-telling behaviour by the bidders. Thus we can design an auction where the bidders will be willing to reveal their true values if asked. We therefore consider the following type of auction. Each bidder is asked to reveal his true value to the seller. A bidder with a high value is given the probability of h of winning the auction and paying H for the object, and a probability of $(1 - h)$ of losing and paying zero. A bidder with a low value is given the probability of k of winning the auction and paying L for the object, and a probability of $(1 - k)$ of losing and paying zero. Thus the values H, L, h and k fully characterise the auction. We shall explain later how the probabilities h and k might be implemented in practice. Next, however, let us examine the constraints on the values which the terms H, L, h and k may take.

The constraints fall into two categories: those implied by the behaviour of the bidders and those implied by simple probability theory. In the first category are the participation constraints and the truth-telling or self-selection constraints of the bidders. The latter may be written as follows:

$$(v_H - H)h \geq (v_H - L)k \tag{11.13}$$

and

$$(v_L - L)k \geq (v_L - H)h \tag{11.14}$$

The first of the above two constraints simply ensures that a bidder with a high value has an incentive to reveal it; the term on the left-hand side is his expected payoff from the auction if he reveals his value truthfully, and the term on the right is his expected payoff if he lies. The second constraint similarly ensures that a bidder with a low value does not lie. Given that the truth-telling constraints are satisfied, bidders will tell the truth and be allocated the appropriate probability of winning the auction and making the payment designed for them by the seller in determining the optimal auction.

The participation constraints are straightforward, since a truth-telling bidder will only participate if he values the object for sale at least as much as he will have to pay for it if he wins the auction. Thus the participation constraints are:

$$v_H \geq H \tag{11.15}$$

and

$$v_L \geq L \tag{11.16}$$

The remaining constraints follow from simple probability theory. In a symmetric auction with two bidders between whom the seller is unable to distinguish, they will both be given an equal chance of winning by the auction design; this chance clearly cannot exceed 0.5. We therefore have:

$$(1 - p)h + pk \leq 0.5 \tag{11.17}$$

where the left-hand side represents the chance that the seller attaches to any bidder winning the auction before the bidders reveal their types – that is, it is the chance that a bidder will prove to be a high type times the associated probability of winning plus the chance that the bidder is a low type times the associated probability of winning.

It is also necessary to constrain the value of h, since a high-type bidder cannot be given a better chance of winning than if he always wins when his opponent is a low type and he is given an equal chance of winning if his opponent is a high type. This means that:

$$h \leq p + (0.5)(1 - p) = (0.5)(p + 1) \tag{11.18}$$

Similarly, a low-type bidder cannot be given a better chance of winning than if he always wins when his opponent is a high type, and he is given an equal chance of winning if his opponent is a low type. This means that:

$$k \leq 1 - (0.5)p \tag{11.19}$$

The optimal auction determines values for H, L, h and k which maximise the expected revenue of the seller which is given by:

$$R = 2[(1 - p)hH + pkL] \tag{11.20}$$

where the term in the square brackets represents the expected payment to the seller from either of the two bidders, and the total expected revenue is simply double this. The first term in the square brackets is the probability that a bidder is of the high type times his expected payment to the seller given that the bidder is a high type, and the second term is the probability that he is a low type times the associated expected payment.

The optimal auction therefore maximises R with respect to H, L, h and k and subject to the constraints given in constraints (11.13) to (11.19). This seems like a very complicated constrained optimisation problem but fortunately it is possible to simplify matters considerably by dealing with the two sets of constraints separately to determine which constraints bind and which do not. Let us therefore examine the 'behavioural' constraints using Figure 11.1 and our example values for v_H and v_L.

Figure 11.1 shows the truth-telling and participation constraints for our example for any feasible values of h and k. The participation constraint for a high-type bidder is that H is no greater than 2, and for the low-type bidder that L is no greater than 1, thus giving, respectively, the heavy vertical and horizontal lines shown. The truth-telling constraint for the high-type bidder gives the upward-sloping heavy line shown passing through the point (2,2) with points on or to the left of it satisfying the constraint. Similarly, truth-telling for the low bidder gives the line shown passing through the point (1,1) where points on or to the right of it satisfy the constraint. The reader may manipulate constraints

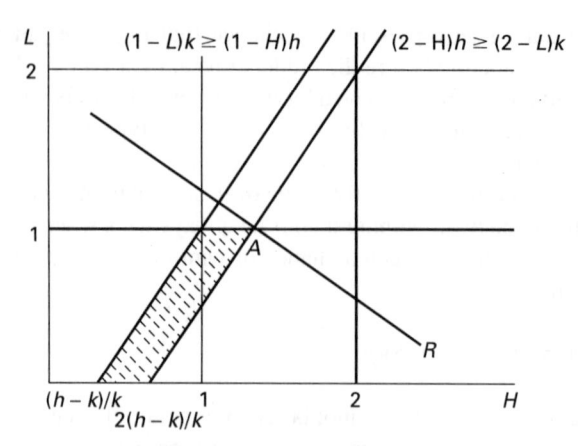

168   *Regulation, Public Procurement and Auctions*

```
L |   (1 − L)k ≥ (1 − H)h      (2 − H)h ≥ (2 − L)k
2 |
  |
  |
1 |                    A
  |
  |                           R
  |_____ H
   (h − k)/k    1          2
        2(h − k)/k
```

Figure 11.1   Truth-telling and participation constraints

(11.13) and (11.14) to derive these last two lines and to show that they both have a common positive slope of h/k. Points on the figure which satisfy the four constraints produce the shaded area.

Manipulation of equation (11.20) to find the slope of an iso-revenue curve in (H, L) space yields a slope of $(p - 1)h/pk$, which is negative. The highest or revenue maximising iso-revenue curve, will therefore touch the shaded area at point A, as shown in Figure 11.1. At this point the truth-telling constraint for the high bidder is binding (holds with equality), as is the participation constraint for the low-value bidder. Hence we may use L equals 1 and k equals $h(2 - H)$ (found using L equals 1 in the high bidder's truth-telling constraint) in equation (11.20) to yield:

$$R = 4[(1 - p)h + k(p - 0.5)] \qquad (11.21)$$

Since equation (11.21) expresses R in terms of the choice variables k and h we may now use it alongside the remaining 'probability' constraints (11.17) to (11.19) in Figure 11.2. Using our example value of p equals 0.5, the constraint (11.17) yields the heavy downward-sloping line $k + h = 1$; points on or below this line satisfy the constraint. Similarly, constraint (11.18) yields the heavy vertical line $h = 0.75$, and constraint (11.19) yields the horizontal line $k = 0.75$, with points on or to the left of the vertical line and points on or below the horizontal line satisfying the constraints. Therefore the shaded area shown represents the set of points capable of satisfying all three constraints.

Using p equals 0.5 in (11.21) yields R equals $2h$ so that the iso-revenue curves in this diagram are vertical. R will therefore be maximised by setting h equal to 0.75, which occurs where the highest achievable iso-revenue curve is coincident with the vertical constraint shown in the figure between points a and b. At point a, k equals 0 and at point b, k equals 0.25 (as determined by the intersection of

Auctions 169

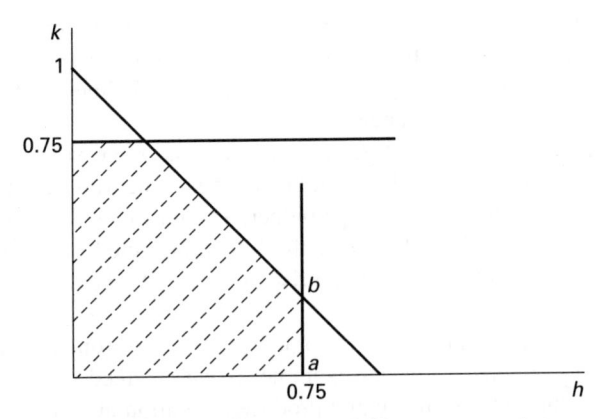

Figure 11.2 *The probability constraints*

the two constraints at b). Hence the optimal value of h is 0.75 and that for k may be anywhere between 0 and 0.25.

Notice that h takes its maximum possible value, which means that a high bidder is guaranteed to win the auction when his opponent is a low bidder, which occurs with probability 0.5, and the high bidder wins with a probability of 0.5 when the other bidder is high, which also occurs with probability 0.5; the overall probability of a high bidder winning is, therefore, 0.75. Using L equals 1, which we derived earlier, and any pair of appropriate h and k values in the binding truth-telling constraint, constraint (11.13), yields the value for H. When k equals 0 (at a), this yields H equals 2 and when k equals 0.25 (at b) this yields H equals 1.66(recurring).

The optimum H values are easily interpreted. When H equals 2, the seller is setting a take-it-or-leave-it price of 2, which is the value of the object held by high bidders. The seller therefore rules out the possibility of a sale to a low bidder and sets k equal to 0. When the value of k is positive this reduces the value of H and means that the seller is no longer setting a take-it-or-leave-it price but is prepared to sell to a low bidder at a value of 1 when both bidders reveal that they are low bidders. Notice that whenever k is positive and L equals 1 the low type of bidder is forced on to his participation constraint, which holds with equality, and does not gain from the auction. On the other hand, for values of H less than 2 (which are set whenever k is positive) the high-type bidder's participation constraint holds with inequality and he expects to gain from participating in the auction although his truth-telling constraint is binding.

The seller is indifferent between the pairs of H and k values determined by points along the line between a and b in Figure 11.2 as may be verified by examining the expected revenue, R, for the pairs of values H equals 2, k equals 0 and H equals 1.66, k equals 0.25. In the former case, R is given by 0.25 times 0 plus 0.75 times 2 (that is probability of both bidders being low types times zero (since, when k is zero, the low types never win the auction) plus the probability of one of the two bidders being a high type times the conditional payment by a high type on winning the auction) which yields 1.5 for the expected revenue and in the latter case it is given by 0.25 times 1 plus 0.75 times 1.66 which also yields 1.5 for the expected revenue.

Notice that, within the set of the optimal values, as the value chosen for k rises it is necessary for H to fall to continue to satisfy the truth-telling constraint in constraint (11.13) for the high-type bidder. The reason why revenue remains constant as k rises and H falls is that the fall in expected revenue from high types as H falls is exactly offset by the rise in expected revenue from low types as k rises.

The optimal auction in this case is not unique but this property depends upon the values for p, v_H and v_L in our example; if we changed the value of p we would remove the indeterminacy and produce a solution at a corner point such as either a or b, but not at points in between the two corners (as the reader will be asked to verify in problem 11.1 below).

The expected revenue from any optimal set of H, h, L and k values in our example is 1.5, which may be compared with the expected revenue from the English auction of 1.25, to show how the optimal auction increases the seller's expected revenue. The increase in revenue compared with the English auction or other standard auction types may be viewed as being caused by the fact that the optimal auction manages to attract as high a bid as possible from a high-type bidder without encouraging him to mimic a low-type bidder (given that the truth-telling constraint (11.13) is satisfied).

The expected revenue from the optimal auction lies below the amount that the seller could achieve if he was able to observe the values held by the bidders, in which case he could always extract 2 from high bidders or 1 from low bidders, giving an expected revenue figure of 0.25 times 1 (for the case where both bidders turn out to be low bidders) plus 0.75 times 2 (for the cases where at least one of the bidders is a high type and pays 2), which yields 1.75. This latter value for expected revenue is, however, unobtainable given the information problem posed by privately-known values.

Notice that the optimal auction is optimal only in the sense of maximising the seller's expected revenue. In the case where k is set to zero there is a positive probability that the object will remain unsold which, assuming that the seller values the good at zero as a necessary condition for him to wish to maximise the expected revenue (rather than to maximise the expected excess of revenue over a positive reservation value) means that there is a positive probability that the good will remain in the seller's hands even though there is a bidder who values it more than the seller. In such a case it is clearly preferable from a social point of view that the good should be sold to the bidder; therefore, the optimal auction with k equal to zero does not produce a Pareto optimal outcome. However, the optimal auction with a positive k value guarantees a sale and is Pareto optimal.

In more general cases than the two-bidder case we have examined it is likely that seller revenue can be increased by using a reserve price above the minimum value held by any possible bidder (this is similar to our case when k is set to zero and H equals 2). In such cases there is, of course, a chance that the object for sale will not be sold and that it will remain in the hands of the would-be seller, who values it less than one of the unsuccessful bidders, and the outcome will clearly not be Pareto optimal. Nevertheless, in practice this might mean only that a sale is deferred, since the seller is likely to bring the object back for sale at a later date.

11.5 Common value auctions and the winner's curse

The results in the previous two sections are based on the extreme assumption that the bidders know their own private values and that these are independent of other bidders' values. This assumption might be appropriate for auctions of some types of good, say second-hand, non-antique furniture. I rule out antique furniture since, although, being a philistine, I may place little private value on a Louis xv black lacquer writing table with gold decoration and chased ormolu mounts, I might be prepared to pay a considerable sum for it in the hope of a profitable sale to someone else at an even higher price. One might argue therefore that antiques are more like goods with a common but unknown value; for instance, all the dealers at an antique auction may simply hope to be able to sell the objects they buy at a profit to private collectors whose private values the dealers can only guess. Similarly, the auction of, say, an offshore oil lease may best be described by the common value model. In this model the value of the lease will be common, but unknown, to all bidders, depending on how many barrels of oil may be extracted from the tract of oilfield being auctioned. Each bidder is likely to have a different estimate of the number of barrels of oil the tract will yield and, hence, to have a different private valuation or estimate of the true value of the lease.

Consider a first-price sealed-bid auction of an offshore oil lease. Assume that the valuation of each bidder is an unbiased estimate of the true common value of the oil lease, so that if bidders participate in many auctions they will each, on average, guess the true common value despite sometimes guessing too high and sometimes too low.

It may seem that risk-neutral bidders in the common value model should behave as those in the private value auction analysed in section 11.3 above, simply replacing their value in the optimal bid formula found there by their valuation. It is likely, however, that if they followed this strategy they would, on average, lose a considerable amount of money.

The optimal bid found in section 11.3 involved shading one's bid below one's value in an attempt to trade off the probability of winning the auction, against other bidders with lower private values, with the surplus to be gained from winning it at a lower bid. In a common value auction, however, it is necessary to shade one's bid, for two reasons. The first is similar to that in private value auctions, with the only difference now being that the other bidders do not have different values but different valuations. The second reason is very different. If all bidders followed the optimal bidding procedure found for the private values model (having substituted his valuation for his value), the winner in the common value model would be the one who held the highest valuation out of a set of valuations, all of which are unbiased. A better estimate (with a smaller variance than any single valuation) of the true common value than any single valuation would be the average of all valuations. Hence the highest valuation tends to overestimate the true value and the winner of the auction would tend to lose money; this effect is known as the winner's curse.

Thus rational bidders in common value auctions shade their bids even more than in the private values case in order to account for the selection bias and try to avoid the winner's curse. Unfortunately, calculating the optimal bid in the common value model is rather difficult so we shall not present any detail here. Perhaps the difficulty of calculating the optimal bid explains the evidence that the winner's curse is difficult to avoid! The interested reader may pursue some of the references given in the next section.

11.6 Recommended reading

The literature on auctions tends to be difficult but the interested reader will find a fairly comprehensive survey in McAfee and McMillan (1987). McAfee and McMillan (1986) analyse the bidding for government contracts as a principal–agent problem. The seminal work on optimal auctions was by Myerson (1981) and our two-bidder, two-value example was based on one found in Binmore (1992, chapter 11). Bulow and Roberts (1989) try to make the topic of optimal auctions simple. The articles by Milgrom, Ashenfelter, Boyes and Happel, and Riley in the symposium on auctions in the *Journal of Economic Perspectives* (1989) are all accessible.

The symposium article by Ashenfelter (1989) injects several notes of realism and jargon familiar to anyone who has experienced real auctions. For example, he points out that auctioneers rarely reveal reserve prices and that in real auctions all objects placed for sale appear to be sold or *knocked down* by the auctioneer. However, it may be that an object has not really been sold despite appearances. In such cases the objects are said to have been *bought in* by the auctioneer, although this does not mean the auctioner will buy it, only that the object did not reach its reserve price. Only after an auction does the auctioneer reveal which objects were, in fact, sold. A likely reason for *buying in* like this is that a seller may prefer to see his object remain unsold and offer it again at a later date when other bidders may bid the price up higher, or even to sell it privately. In such cases the highest actual bid offered for the object serves as a useful signal to the seller, since it gives him some information about the sort of price to expect.

11.7 Problems

Problem 11.1
In Section 11.3 above, Ben chose to hold an English auction rather than a first-price sealed-bid auction, in order to avoid collusion between the auctioneer and a bidder. Show, using examples for a first-price sealed-bid auction, that:

(a) It is possible for Ben to gain from such collusion.
(b) It is possible for Ben to lose from such collusion.

On balance, do you think it is more likely that Ben would gain or lose from such collusion?

Problem 11.2
Consider the optimal auction problem of section 11.4 above and derive the optimal auction rules for a general value of p. Show that the value of p equals 0.5 used in section 11.4 was special and that no other feasible value leads to multiple optimal values for the auction parameters. Compare and contrast the optimal auction rules for values of p greater than 0.5 with those for values of p less than 0.5.

Problem 11.3[9]
Eric wants to sell a ticket to a football game. He values it at £40. He has two potential buyers, Fred and George. He only knows that Fred and George have values that are independently and uniformly distributed between £40 and £50. Similarly, neither Fred nor George knows the other's value, only that it is independent of his own value and is uniformly distributed between £40 and £50. Let F be Fred's value and G be George's value, in pounds. Eric decides to hold an auction.

(a) If Eric decides to hold an English auction, then the expected net return to Fred is $R_F = qF - P$, where q is the probability of winning the auction and P is the *expected* payment. The probability that Fred wins is prob$[G < F] = 0.1(F - 40)$, since G is uniformly distributed between 40 and 50. Since in an English auction the winner pays the second highest value, the expected payment is given by the following definite integral: $P = \int_{40}^{F} (0.1\, G)\, dG$.

  (i) Calculate the probability that Fred wins if his value $F = 40$; if $F = 45$; if $F = 46$; if $F = 50$.
  (ii) Calculate Fred's expected payment, P, as a function of his value.
  (iii) Calculate Fred's expected payment *conditional on winning* the auction. What is his expected payment conditional on winning if $F = 46$? What is his expected net return if $F = 46$?

(b) Suppose Eric decides to hold a first-price sealed-bid auction. Suppose also that Fred and George adopt a linear bidding strategy, that is b_i equals $a_i + c_i i$ for i equals F or G, where b_i is the bid and a_i and c_i are constants. The expected net return to Fred is now $R_F = q(F - b_F)$. The probability that Fred wins is now $q = \text{prob}[b_G < b_F] = \text{prob}[a_G + c_G G < b_F] = \text{prob}[G < (b_F - a_G)/c_G]$.

(i) Calculate the probability that Fred wins as a function of his bid b_F and George's bidding parameters a_G and c_G.

(ii) If Fred chooses his bid to maximise his expected net return, show that his optimal bid as a function of his value F and George's bidding parameters is $b_F = (0.5)(a_G + 40c_G) + (0.5)F$.

(iii) From the previous part we have $a_F = (0.5)(a_G + 40c_G)$ and $c_F = (0.5)$. Show that in a symmetric equilibrium $a_F = a_G = 20$, and $c_F = c_G = 0.5$.

(iv) If $F = 46$, calculate Fred's optimum bid, the probability that he wins, and his expected net return. Compare your answers here to those for part (iii) of part (a) above and relate the results to the revenue equivalence theorem.

Notes

Introduction

1. It is mirrored closely in the literature on time inconsistency (see, for example, Hillier and Malcomson, 1984) and contract renegotiation (see, for example, Fudenberg and Tirole, 1990).

Chapter 1   Asymmetric Information in the Market for Investment Finance

1. Clearly, in the selection and hidden actions problems, respectively, the bank may be able to distinguish between different borrowers or observe the use to which borrowers put funds loaned to them if it incurs costs, but for simplicity we deal with the extreme cases where the bank has no way, even at cost to itself, of distinguishing between borrowers or observing the use made of funds.
2. Readers interested in game theory are recommended to examine any of the many good introductory texts available, such as Binmore (1992).

Chapter 2   Investment Finance and the Selection Problem

1. Clearly, since, from the participation constraint, only entrepreneurs with projects for which R_i^s is greater than $(1 + r)K$ apply for funds, then all successful projects are able to repay the loan.
2. ρ_i is simply calculated using the formula:
$$\rho_i = [K(1 + rp_i) - K]/K = rp_i.$$
3. Remember that we assume that even though the participation constraint is met with equality at an interest rate of 30 per cent for entrepreneurs with type 1 projects, they continue to apply for loans. This assumption merely simplifies the analysis by giving us a clear cut-off point rather than having to talk about such entrepreneurs seeking funds as long as the interest rate was less than 30 per cent. We make a similar assumption for entrepreneurs with type 2 projects when the interest rate is 40 per cent.
4. The case where the horizontal line representing d^* cuts through the discontinuous part of the $\rho - r$ relationship is left for the reader to consider as an exercise.
5. The latter result, of entrepreneurs with type 1 projects withdrawing from the market, is reminiscent of the adverse selection result in the seminal paper by Akerlof (1970). Akerlof set up a model where high-quality second-hand cars were not offered for

176   *Notes*

      sale and only low-quality ones, called lemons, were offered. The term 'adverse selection' is more appropriate for Akerlof's case than ours, however, since in an objective sense type 1 and type 2 projects are equally good in our example: they both cost 100 and both yield an expected gross return of 120.

6. The actual returns to banks and entrepreneurs depends upon the type of contract. For instance, under a share arrangement, an entrepreneur whose project fails receives a positive payoff, while under a credit arrangement he receives nothing; therefore, risk averse entrepreneurs would prefer a share arrangement (see Chapter 5 below).
7. We can assume that entrepreneurs wish to sell only enough shares to fund their project, rather than sell 100% of the project, if we assume that they value the expected return per share of 1.2 (that is, 120/100) more than the price of shares, which is 1.0169; that is, we assume that if the entrepreneurs had funds they would be happy to receive a lower rate of return than that required by the suppliers of funds.
8. Entrepreneurs with type 2 projects able to borrow at a quoted loan rate of 36 per cent would not be willing to sell shares for a price below 1.0169 so this would, strictly, require us to redraw Figure 2.4 to allow for different supply curves for shares in type 1 and type 2 projects.
9. The interested reader may like to ponder the implications for efficiency of 'limited liability' for the case where project returns may be negative as a result of damage caused to third parties, but where the owners of equity are not liable to pay compensation for this damage.
10. We assume that the average rate of return at B is not affected by some entrepreneurs with type 2 projects getting funds at r_C, since we assume they first of all apply for loans at the lower interest rate, so that loans at the lower interest rate have the population average chance of success.
11. The reader may note that the suppliers of funds now produce the demand curve for shares rather than the supply curve for credit and the entrepreneurs produce the supply curve for shares rather than the demand curve for credit. Nothing substantial hinges on the switching of the supply and demand terminology; for instance, in the credit market case, we could have imagined entrepreneurs to be offering to supply promises to pay which were being demanded by the suppliers of credit. It seems natural, however, to imagine entrepreneurs to be demanders of loans but suppliers of shares.
12. We assume that entrepreneurs either sell enough shares to fund their projects, or none at all. Hence the equilibrium shows some, but not all, projects being funded rather than partial funding of all projects.

Chapter 3   Investment Finance and the Hidden Action Problem

1. Obviously, banks can to some extent monitor what borrowers do with loans, but this may be costly and imperfect. Assuming that banks are completely unable to observe the act of investment greatly simplifies matters.
2. In the case of the hidden action problem it is necessary that the participation constraint for the entrepreneur, as discussed in the previous chapter, is satisfied, but it is usually more important to examine the incentive compatibility constraint, and the participation constraint is often not considered explicitly.
3. Strictly speaking, the assumption of epsilon altruism implies that we should add the amount epsilon, ϵ, to the left-hand side of equation (3.2).
4. This case is often known simply as *moral hazard*. We use the longer term to differentiate it from the case of moral hazard with hidden information to be discussed in the

next chapter. When it is clear which case we are discussing, however, we shall often use the shorter term to refer to either case.
5. The seminal reference on moral hazard is Arrow (1963).
6. In the short run, entrepreneurs may need to induce banks to switch to equity finance by selling shares at a price slightly below 1.11, thus offering them a higher expected effective return than the credit market. In the long run, however, competition between banks would drive the share price up to 1.11.
7. A Pareto improvement is produced by a move from one situation to another where nobody loses and somebody gains from the switch.
8. We assume that it is not possible to observe where the entrepreneur invests but that it is possible to observe whether he invests or not; otherwise, the entrepreneur would not invest at all but simply keep all the funds made available to him.
9. This is an application of the 'sufficient statistic condition', since the punishment is based on evidence that is sufficient to indicate the entrepreneur's action and is only incidentally based on payoffs (the entrepreneur is punished for a payoff of 100, which reveals his action, but not for the lower payoff of zero, which does not reveal his action). See Holmstrom, (1979). Since the projects in the example given in section 3.2 also produce different payoffs if successful, the reader may notice that the type of legal solution presented here could also be applied in that case; although, as section 3.3 showed, a switch to equity finance would solve the hidden action problem for that example without need of recourse to the legal system.
10. The possibility that everyone can gain as a result of this punishment policy explains why individuals may be willing, in extreme cases, to provide hostages to verify that punishment can be carried out if necessary. The provision of hostages allows gains to be made (it would prevent market collapse in our example) and, as long as the hostage provider does not succumb to moral hazard, the hostages will not be harmed.
11. In other words, the payoff distributions may be said to have different supports. Reward schemes based on taking advantage of different distributions like this are often known as 'shifting support schemes'.
12. The expected return to the entrepreneur can most easily be calculated by noting that it equals the gross expected return to the project less that part of it expected to go to the suppliers of funds. Thus, if the suppliers of funds of 100 want an expected return of 115 to yield them a rate of return of 15 per cent, then the entrepreneur expects a return of 15 from type 2 projects (130–115) and 18.33 from type 1 projects (133.33–115).

Chapter 4   Investment Finance and the Costly State Verification Problem

1. If the announced return is the maximum of 200 then the bank need not monitor, since the actual return could not exceed the maximum. Allowing for this, however, makes no difference to the arguments in the text.
2. We assume throughout the remainder of this chapter that banks commit themselves to monitor whenever an entrepreneur declares a payoff less than some critical value, but it is worth noting that in some cases it may be sufficient to induce truth-telling to let the entrepreneur know that he will face a probability of being monitored if he declares a payoff below the critical value. Randomising monitoring in this way is useful because it reduces the amount of monitoring carried out and saves on monitoring costs. For an interesting, but difficult, paper demonstrating this result see Mookherjee and Png (1989).
3. The diagram shows the rays starting from the origin. Since payoffs lie between *a* and *b* in our example, we would only be interested in that part of the diagram for which R lies between these two values.

4. The interest rate of 34.32 per cent was calculated using calculus and the properties of the uniform distribution.
5. In general, the maximum payment must be set at a level of b minus c, which yields 190 in our case, or an interest rate of 90 per cent. The critical rate of 90 per cent for our example is, of course, a result of the assumptions made in that example and should not be taken to indicate that the critical rate is so high as to be unimportant in practice.
6. The reader may derive this curve by using a four-part diagram similar to that drawn in Figure 2.8 for Problem 2.1. The loan rate at which the $\rho - r$ relationship begins to turn down in that figure is the critical value discussed in the text.
7. For the sake of simplicity we do not deal in detail with the labour market. The underlying idea, however, is that entrepreneurs pay a market clearing wage to attract labour to work with the capital invested in their project. Thus the real wage is equal to the marginal product of labour. The real wage depends upon the productivity of capital and entrepreneurs are assumed to form rational expectations of what this real wage will be when deciding whether or not to fund their project. These expectations will be correct unless there is a shock which entrepreneurs could not predict; such shocks are introduced in the text below when we consider productivity shocks. For a more formal treatment of the labour market in this type of model, see Bernanke and Gertler (1989) or Hillier and Worrall (1995).
8. A similar idea would be to make the investment cost common across entrepreneurs but allow the project returns to be drawn from different distributions. All that is important for the analysis is that some entrepreneurs' projects are better than others.

Chapter 5   Insurance and Risk Aversion

1. We assume throughout that the first derivative, $U'(W)$, is positive; that is, we assume that individuals prefer more wealth to less. This seems to be a reasonable assumption.
2. Alternatively, we could assume that insurance companies have so many customers that the aggregate risk they face is zero; that is, a proportion p of their customers require compensation and a proportion $(1-p)$ do not, so that they can calculate their returns with certainty in the aggregate and act accordingly.
3. The gap between $E(W)$ and S represents the maximum increase in premium above pC, for compensation of Y equal to C, which an insurance company in the real world could charge to cover running costs of the business and non-zero normal profits.
4. For a formal proof of this statement, see the discussion of the fair odds line in the next section.
5. Solving Problem 5.2 below will make this point clear to the reader.

Chapter 6   Insurance and the Hidden Action Problem

1. A similar problem would occur if the insurance company was prevented by law from making contracts contingent in this way. Our assumption, however, is that they are prevented not by law but by the prohibitive costs of finding out or proving that the actions taken by the individual were such as to warrant a change in premium or compensation.
2. If the insuree was allowed to choose any point along FF' while exercising the low level of care, he would choose more-than-full insurance at a point at which a low-care indifference curve is tangential to FF'; this choice maximises his expected utility

Notes 179

along FF'. Obviously, such a choice would involve losses for the insurer and would not be allowed in equilibrium.
3. However, one possible reason why more businesses report fires to their premises during recessions might be that the true value of their premises has fallen below their insurance value as a result of the recession and provided an incentive for the owners to take less care to avoid fires – or even, in the extreme, to commit arson.

Chapter 7   Insurance and the Selection Problem

1. Obviously, in the real world, insurance companies attempt to place their customers into different risk categories, but within any given category there will be different types of customer and the analysis to be given in the text may be applied to a given category of customers.
2. Since the slope of the indifference curve for the safe customer through the market average fair odds line where it cuts the full-insurance line is steeper than the slope of the market average fair odds line, it follows that a tangency point between an indifference curve for the safe customer and the market average fair odds line will occur below the full-insurance line. On the other hand, for a risky customer, the slope of the indifference curve is less steep than the slope of the market average fair odds line where they cut the full-insurance line, so that he would prefer more than full insurance if possible. The reader may verify these arguments by drawing a diagram.
3. See Chapter 5 and Problem 5.2 to gain an understanding of this point. Notice also that such gambles offer unfair odds to the risky customers and so would not be chosen by them in preference to the endowment point.
4. Of course, it is also necessary that the safe customers prefer the partial insurance contract at R to full insurance at K, but this is guaranteed by the nature of the problem. The safe customer would prefer partial insurance to full insurance along HH' and would prefer point R to any point on HH' as the reader may verify by remembering that for any point in the diagram the indifference curve for the safe customer is steeper than that for the risky customer.
5. Both concepts are in a sense based upon rivals' reactions and both could easily have been called Reactive equilibrium. It may therefore have been better to have called the Reactive equilibrium after its originator, Riley. The terminology is by now, however, quite standard.

Chapter 8   The Selection Problem and Education

1. These tests are based upon the *intuitive criterion* of Cho and Kreps (1987). See Binmore (1992, ch. 11) for a clear presentation and a criticism of this criterion; the criticism, however, would not seem to apply to the use of the communication tests in the context of the material in this chapter.

Chapter 9   The Hidden Action Problem and Efficiency Wages

1. Some economists might even ask, 'Does the labour market fail to clear?', but we shall assume that unemployment does exist and needs to be explained.

180   Notes

2. Ford may have been inspired by the earlier experience of Percival Perry in Manchester. Perry had explained his decision to increase pay when Ford visited England in 1912.
3. When Henry Ford introduced the $5 day it was paid only to workers who had had a long enough period of employment at the Ford plant to qualify, and the workers feared that they would be fired before they qualified. Ford took great steps to acquire a reputation for honesty in this regard so as to convince the workers that they could achieve such a high rate of pay.
4. Saint-Paul (1995), however, shows how the shirking-model may be extended to explain persistence or cyclical effects.

Chapter 10   Regulation and Procurement

1. Until the 1980s such companies in the United Kingdom were nationalised or held in public ownership as a way of controlling them. Although the privatisation programme has returned many of these companies to the private sector, this change has been accompanied by setting up regulatory frameworks to control the monopoly powers of the newly-privatised companies.
2. The reader requiring a discussion of the calculation of consumer surplus is referred to any standard intermediate microeconomics text.
3. If left to maximise profits without regulation the monopolist would, of course, choose to raise the price to that level at which the marginal cost, C, equals marginal revenue.
4. These technical assumptions simply guarantee that the optimal value for e is positive and less than R.
5. Assuming, of course, that P^* is less than S.
6. The reader may be interested to consider what would happen if we introduced only one of the two information problems and not both. The result is that the authority should be able to achieve the full-information outcome as long as it can observe two of N, R and e, since from any two it can calculate the third.
7. Obviously, the authority cannot offer both prices and ask the firm to choose the lower one if it is a type L firm, since such a firm rationally would choose to accept the higher price level.
8. We are assuming that the authority's welfare function is given by $S - P$ and that it does not attach any weight in that function to the excess profits made by the type-L firm when the price is set at P_H. It is not difficult to introduce a positive weight on firm profits into the welfare function; see the references in Section 10.4.
9. Of course, the purchaser might gain from reneging in this way once but he would then lose his reputation for honouring contracts and then lose in any future deals. Thus purchasers can benefit from having a good reputation or from a legal system that guarantees they will be forced to abide by accepted contracts.
10. Such renegotiation is not possible when facing a low-cost producer, since no renegotiation could improve the welfare of both parties.
11. The interested reader is advised to read Chapter 11 on auctions and then to pursue the recommended reading for the present chapter for applications of the ideas to purchasing problems.

Chapter 11   Auctions

1 It may be the case in some auctions – for example, a government auctioning the rights to a television broadcast channel – that the seller is concerned about details of

the bids other than price. However, we assume in the text that the seller is concerned only with the price dimension of bids.
2. In the text we consider only single object auctions but sometimes multiple units are auctioned simultaneously. For example, a government auctioning debt might adopt a rule such that it sells its debt in the amount demanded by the highest bidders by working down the list of bidders until all the debt on auction is demanded and is all sold at the lowest winning bid. The reader may like to ponder whether this auction should be categorised as discriminatory or non-discriminatory.
3. Non-refundable auctions do occur. For instance, sometimes all bidders at an auction organised to raise funds for a charity are required to pay their bid. While such auctions may not be of much practical interest it should be noted that they are analytically similar to other more important topics. For example, in a race to develop a new product all participating companies pay their 'bid' – that is, their expenditure on R&D – although the prize may go only to the winner; or, in a contest to host the Olympic Games, all the competing cities incur non-refundable expenditure when attempting to impress the Olympic Committee. Clearly, such contests are very similar to non-refundable auctions, although the 'bids' of losers may simply be dissipated rather than accrue to the auctioneer or seller.
4. Since we assume that v is distributed according to a continuous uniform distribution, we assume that the possibility of a tie is zero.
5. This does not prove that other solutions would not also satisfy equation (11.8) but the reader may be assured that the solution we find is, indeed, the unique solution.
6. Using a result on order statistics.
7. Again using a result on order statistics.
8. With possibly severe and illegal punishments for members who are caught cheating.
9. I thank Tim Worrall for this question.

Bibliography

Akerlof, G. (1970) 'The Market for Lemons: Quality Uncertainty and the Market Mechanism', *Quarterly Journal of Economics*, vol. 84, pp. 488–500.
Akerlof, G. (1982) 'Labor Contracts as Partial Gift Exchange', *Quarterly Journal of Economics*, vol. 97, pp. 543–69.
Arrow, K. J. (1963) 'Uncertainty and the Welfare Economics of Medical Care', *American Economic Review*, vol. 53, pp. 941–73.
Ashenfelter, O. (1989) 'How Auctions Work for Wine and Art', *Journal of Economic Perspectives*, vol. 3, pp. 23–36.
Baron, D. P. and R. B. Myerson (1982) 'Regulating a Monopolist with Unknown Costs', *Econometrica*, vol. 50, pp. 911–30.
Bernanke, B. S. and M. Gertler (1989) 'Agency Costs, Net Worth, and Business Fluctuations', *American Economic Review*, vol. 79, pp. 14–31.
Bester, H. (1985) 'Screening vs. Rationing in Credit Markets with Imperfect Information', *American Economic Review*, vol. 75, pp. 850–5.
Binmore, K. (1992) *Fun and Games: A Text on Game Theory* (Lexington, Mass.: D. C. Heath).
Boyes, W. J. and S. K. Happel (1989) 'Auctions as an Allocation Mechanism in Academia: The Case of Faculty Offices', *Journal of Economic Perspectives*, vol. 3, pp. 41–50.
Bulow, J. and J. Roberts (1989) 'The Simple Economics of Optimal Auctions', *Journal of Political Economy*, vol. 97, pp. 1060–90.
Cho, I.-K. and D. M. Kreps (1987) 'Signalling Games and Stable Equilibria', *Quarterly Journal of Economics*, vol. 102, pp. 179–221.
Fazzari, S. M. and B. C. Petersen (1993) 'Working Capital and Fixed Investment: New Evidence on Financing Constraints', *Rand Journal of Economics*, vol. 24, pp. 328–42.
Fudenberg, D. and J. Tirole (1990) 'Moral Hazard and Renegotiation in Agency Contracts', *Econometrica*, vol. 58, pp. 1279–319.
Grossman, S. and O. Hart (1983) 'An Analysis of the Principal Agent Problem', *Econometrica*, vol. 52, pp. 1–45.
Helm, D. (1994) 'British Utility Regulation: Theory, Practice and Reform', *Oxford Review of Economic Policy*, vol. 10, pp. 17–39.
Hillier, B. and M. V. Ibrahimo (1992) 'The Performance of Credit Markets under Asymmetric Information about Project Means and Variances', *Journal of Economic Studies*, vol. 19, pp. 3–17.
Hillier, B. and M. V. Ibrahimo (1993) 'Asymmetric Information and Models of Credit Rationing', *Bulletin of Economic Research*, vol. 45, pp. 271–304.
Hillier, B. and J. M. Malcomson (1984) 'Dynamic Inconsistency, Rational Expectations and Government Economic Policy', *Econometrica*, vol. 52, pp. 1437–51.
Hillier, B. and T. Worrall (1994) 'The Welfare Implications of Costly Monitoring in the Credit Market', *Economic Journal*, vol. 104, pp. 350–62.

Bibliography

Hillier, B. and T. Worrall (1995) 'Asymmetric Information, Investment Finance and Real Business Cycles', in H. Dixon and N. Rankin (eds), *The New Macroeconomics* (Cambridge University Press).

Hirshleifer, J. and J. G. Riley (1992) *The Analytics of Uncertainty and Information* (Cambridge University Press).

Holmstrom, B. (1979) 'Moral Hazard and Observability', *Bell Journal of Economics*, vol. 13, pp. 74–91.

Laffont, J. J. and J. Tirole (1986) 'Using Cost Observation to Regulate Firms', *Journal of Political Economy*, vol. 94, pp. 614–41.

Laffont, J. J. and J. Tirole (1993) *A Theory of Incentives in Procurement and Regulation* (Cambridge, Mass.: MIT Press).

Layard, R. and G. Psacharopoulos (1974) 'The Screening Hypothesis and the Returns to Education', *Journal of Political Economy*, vol. 82, pp. 985–98.

McAfee, R. P. and J. McMillan (1986) 'Bidding for Contracts: A Principal–Agent Problem', *Rand Journal of Economics*, vol. 17, pp. 326–38.

McAfee, R. P. and J. McMillan (1987) 'Auctions and Bidding', *Journal of Economic Literature*, vol. 25, pp. 699–738.

Milgrom, P. (1989) 'Auctions and Bidding: A Primer', *Journal of Economic Perspectives*, vol. 3, pp. 3–21.

Mookherjee, D. and Png, I. (1989) 'Optimal Auditing, Insurance, and Redistribution', *Quarterly Journal of Economics*, vol. 104, pp. 399–415.

Myerson, R. (1981) 'Optimal Auction Design', *Mathematics of Operations Research*, vol. 5, pp. 58–73.

Neumann J. von and O. Morgenstern (1944) *Theory of Games and Economic Behaviour* (Princeton, NJ: Princeton University Press).

Raff, D. M. G. and L. H. Summers, (1987) 'Did Henry Ford Pay Efficiency Wages?', *Journal of Labor Economics*, vol. 5, pp. 57–86.

Riley, J. G. (1979) 'Informational Equilibrium', *Econometrica*, vol. 47, pp. 331–59.

Riley, J. G. (1989) 'Expected Revenue from Open and Sealed Bid Auctions', *Journal of Economic Perspectives*, vol. 3, pp. 41–50.

Rothschild, M. and J. E. Stiglitz (1976) 'Equilibrium in Competitive Insurance Markets: An Essay on the Economics of Imperfect Information', *Quarterly Journal of Economics*, vol. 90, pp. 629–50.

Saint-Paul, G. (1995) 'Efficiency Wages as a Persistence Mechanism', in H. Dixon and N. Rankin (eds) *The New Macroeconomics* (Cambridge University Press).

Shapiro, C. and J. E. Stiglitz (1984) 'Equilibrium Unemployment as a Worker Discipline Device', *American Economic Review*, vol. 74, pp. 433–44.

Spence, A. M. (1973) 'Job Market Signalling', *Quarterly Journal of Economics*, vol. 87, pp. 355–74.

Stiglitz, J. E. (1977) 'Monopoly, Non-linear Pricing and Imperfect Information: The Insurance Market', *Review of Economic Studies*, vol. 44, pp. 407–30.

Stiglitz, J. E. and A. Weiss (1981) 'Credit Rationing in Markets with Imperfect Information', *American Economic Review*, vol. 71, pp. 393–410.

Varian, H. R. (1992) *Microeconomic Analysis* (3rd edn) (New York: W. W. Norton).

Vickers, J. and G. K. Yarrow (1988) *Privatisation: An Economic Analysis* (Cambridge, Mass.: MIT Press).

Vickrey, W. (1961) 'Counterspeculation, Auctions, and Competitive Sealed Tenders', *Journal of Finance*, vol. 16, pp. 8–37.

Williamson, S. D. (1986) 'Costly Monitoring, Financial Intermediation, and Equilibrium Credit Rationing', *Journal of Monetary Economics*, vol. 18, pp. 159–79.

Wilson, C. (1980) 'The Nature of Equilibrium in Markets with Adverse Selection', *Bell Journal of Economics*, vol. 11, pp. 108–30.

Yellen, J. L. (1984) 'Efficiency-Wage Models of Unemployment', *American Economic Review*, vol. 74, pp. 200–5.

Index

absenteeism 129–30
adverse selection 12, 19, 32, 44, 98, 102–3
agency problem 5–6
Akerlof, G. 33, 130, 135
Ashenfelter, O. 172
auction 153–74
  bid price/probability trade-off 159–60
  bidders 153
  buyers 153
  buying in 172
  collusion 163
  common-value 154, 155, 171–2
  discriminatory/non-discriminatory 156
  Dutch 155–6, 157–65
  English (first-price open-cry) 155, 157–65
  expected and actual revenue 162
  expected revenue maximization 165, 170
  first-price sealed-bid 155, 157–65, 171
  optimal 153, 163–4, 165–71
  optimal bid 171–2
  optimal bidding strategy 161
  Pareto optimal 161–2, 170
  private-value 154, 156–65
  refundable/non-refundable 156
  reserve price 157, 164
  ring 162–3
  second-price sealed-bid 155, 157–65
  seller's 153
  symmetric independent private values (SIPV) model 154, 164, 165
  see also revenue equivalence theorem

banking
  competition 10–11, 61–4
  optimal financial intermediation 64
  received returns 11–13, 14–16
Baron, D. P. 151

Bayes' Rule 122
Bernanke, B. S. 74
Bester, H. 33
Binmore, K. 172
Boyes, W. J. 172
Bulow, J. 172
business cycles
  with asymmetric information 68–74
  and costly state verification 70–1
  and credit market 67–74
  with full information 67–8

care levels 91, 92
certainty line 86
cherry picking contracts 105
collusion 163
communication tests 124–5
compensation 81
  deductible 95
  partial 95
competition, in banking 10–11, 61–4
competition or zero profit constraint 11
constraint
  competition or zero profit 11
  incentive compatibility 37, 43–4, 47, 55, 95–6, 148–9
  no-shirking 132–4
  participation 10, 18, 148, 166, 168
  probability 166–7, 168–9
  truth-telling or self-selection 60–1, 106, 166, 167–8
contract
  cherry picking 105
  cost-plus 144–6, 149–50
  cream skimming 105, 109
  critical value 59–60, 61
  fixed-price 144–6, 149
  insurance 81
  maximum payment 60–1
  pooling 98, 100–5, 109, 121

184

Index

renegotiation 150
  separating 104–10, 119–21, 125, 147, 150
  standard debt 8, 9–10, 32, 58, 61–4
correlated-value case 154
cost
  estimate 142
  function 143
  unit 140
cost-plus contract 144–6, 149–50
costly state verification 4–5, 57–74
  and business cycles 70–1
  and rationing 65–7
cream skimming contracts 105, 109
credit finance
  compared with equity finance 16–22, 28–32, 53–6
  rationing 22–32, 50–3, 65–7
  see also loans
credit market
  asymmetric information 15–16, 19, 52–3
  and business cycles 67–74
  equilibrium 14–16, 51–3
  and hidden action 36–42
  and hidden information 57–67
  inefficiency 32
  Pareto efficient 21
  replaced by equity market 42–3, 55
  selection problem 7–16, 24–7, 103
critical value contract 59–60, 61

debt contract *see* standard debt contact
defection 109
demand
  for funds 9–10, 37–9, 65
  for insurance 80–2
  and supply interdependency 28
deposit rate 11
deposits, supply 13–14, 22, 50
dissipative externality 16, 110, 121

education
  indictor of productivity 118–26
  return to 126
  and screening 115–22
  and selection 115–27
  for signalling 122–5
efficiency 30
  parameter 143–50
  see also Pareto efficiency
efficiency wage models 128, 130
effort 149
endowment point 86

epsilon altruism 38–9, 95, 106, 107, 147
epsilon truthfulness 60
equilibrium
  Bayesian 122–4
  communication-proof 124–5
  credit market 14–16, 51–3
  equity market 18–19, 20, 28–32
  insurance market 94, 99–100
  labour market 116, 134
  loan market 24–7, 40–2
  and market inefficiency 26
  pooling 110, 121–2
  reactive 109–10, 121
  separating 98, 110, 121, 124
  Wilson 109–10, 121
  see also Nash equilibrium
equity finance 28–32
  compared with credit finance 17–22, 30–1
  and hidden action 42–4, 53–6
  rationing 56
  and selection problem 16–22
equity market
  equilibrium 18–19, 20, 28–32
  full information 45–6
  Nash equilibrium 31, 44, 54, 55
  Pareto efficient 21, 44, 55
  Pareto improvement 43
  replaces credit market 42–3, 55
excess 95
expected returns 8
  equation 37, 43–4, 47, 68, 69
  and moral hazard 39–40
expected revenue theorem 161
externality, dissipative 16, 110, 121

fair odds 83–5, 100–1
  line 86–7, 91
  premium 105–6
  see also auction
favourable selection 32
fixed-price contract 144–5, 146, 149
Ford, Henry 129, 135
fraud 49

gambling 80, 83
game theory 5, 20
Gertler, M. 74
gift-exchange models 130
Grossman, S. 96

Happel, S. K. 172
Hart, O. 96
Helm, D. 151

Index

hidden action 4, 36–56
  credit market 37–42
  and demand for funds 37–9
  and efficiency wages 128–35
  and equity finance 42–4, 53–6
  and insurance 90–7
  labour market 130–1, 132–5
  with moral hazard 39, 44, 90–7
  procurement 143–50
  and project returns 37
  and supply of: deposits 50; funds 39; loans 50–1
hidden information
  and credit market 57–67
  with moral hazard 58
  procurement 143–50
Hillier, B. 33, 74
Hirshleifer, J. 89, 124

Ibrahimo, M. V. 33
incentive compatibility 150
  constraint 37, 43–4, 47, 55, 95–6, 148–9
  solution 147
incentive mechanism 38
indifference curves
  break-even 145–7
  insurance contract 87–8, 91–5
  labour market 117
  single-crossing property 100, 117
individual rationality constraint 10
inefficiency see adverse selection
insurance
  compensation 81, 95
  contract 81
  demand for 80–2
  fair odds 83–5, 86–7, 91, 100–1, 105–6
  favourable odds 83
  full 82, 83–5
  and hidden action problem 90–7
  more-than-full 85
  partial 84, 92, 105–6
  premium 81, 101–2, 105–6
  and risk aversion 80–8
  screening model 118
  in selection problem 98–112
  supply 82–3
  unfair odds 83, 88
insurance company, defector 109
insurance customer
  safe and risky 98
  self-selecting 105

insurance market
  asymmetrical information 100–4
  equilibrium 94, 99–100
insurer 82
interest rate
  charged on loans 8
  effective 11
  incentive mechanism 38
  selection mechanism 10, 12
  see also deposit rate; quoted loan rate
investment finance
  demand for funds 9–10
  equilibrium 14–16, 24–7
  project returns 7–8
  and selection problem 7–35
  supply of deposits and loans 13–14, 22–4
  see also banking; credit finance; equity finance; loans

labour, turnover 129–30
labour market
  with asymmetrical information 118–26
  discrimination 130
  dual or multiple 131
  equilibrium 116, 134
  full information 116–17
  indifference curves 117
  selection problem 132
labour productivity 71–2
Laffont, J. J. 151
Layard, R. 126
loans
  market equilibrium 24–7, 40–2
  supply 13–14, 22–4, 50–1, 65
  see also credit finance

McAfee, R. P. 172
McMillan, J. 172
marginal rate of substitution 87
marginal rate of transformation 88
market
  collapse 44–50; legal solution 48–9; and moral hazard 46–8
  conventional 27–8
  failure 26
maximum payment contract 60–1
Milgrom, P. 172
monitoring 49–50, 58–9, 64, 131, 133
monopoly
  unit costs 140
  regulation see regulation
Mookherjee, D. 74

moral hazard
  defined   39
  and expected returns   39–40
  with hidden action   39, 44, 90–7
  with hidden information   58
  and market collapse   46–8
Morgenstern, O.   88
Myerson, R. B.   151, 172

Nash equilibrium   98
  defined   20
  equity market   31, 44, 54, 55
  pooling contract   104–5
  separating contracts   105, 107–8, 119–21, 125
Neumann, J. von   88
no-shirking constraint   132–4
nutrition   129

optimising behaviour   128

Pareto efficiency   21, 44, 55
Pareto improvement   20, 43
Pareto optimum   170
participation constraints   10, 18, 148, 166, 168
perfect Bayesian equilibrium   122–4
Perry, Percival   129, 135
Png, I.   74
pooling case   98
pooling contract   98, 100–5, 109, 121
pooling equilibrium   110, 121–2
principal-agency problem   5–6
probability constraints   166–7, 168–9
procurement
  full information   143–4, 146
  hidden action and hidden information   143–50
productivity   71–2
  education as indicator   118–26
  and wages   115–16
profits
  supernormal   109, 120, 141
  zero supernormal   62, 63
Psacharopoulos, G.   126

quality   27
quoted loan rate   8, 11
  see also interest rate

Raff, D. M. G.   135
rationing
  and costly state verification   65–7
  credit finance   22–32, 50

equity finance   56
reaction   109
reactive equilibrium   109–10, 121
regulation
  asymmetric information   141–2
  choice of contracts   142
  cost estimates   142
  full information   140–1
  and social welfare   139–42
returns   7–8
  banking   11–13, 14–16
  to education   126
  expected   8, 37, 39–40, 43–4, 47, 68–9
  gross   8, 12
  and hidden action   37
Revelation Principle   60–1, 165–6
revenue equivalence theorem   153, 157–65
  assumptions   164–5
  see also auction
Riley, J. G.   89, 111, 124, 172
risk
  attitudes towards   77–80
  aversion   77–89, 98; described   82
  category pooling   100–4
  loving   80
  neutral   9, 77, 79
  under full information   99–100
Roberts, J.   172
Rothschild, M.   110

screening
  compared with signalling   117–18
  and education   115–22
  insurance model   118
secondary sector   131
selection mechanism   10, 12
selection problem   3–4, 32, 121
  credit market   7–16, 24–7, 103
  and education   115–27
  efficiency wage models   130
  and equity finance   16–22
  in insurance   98–112
  and investment finance   7–35
  labour market   132
  see also adverse selection; favourable selection
self-interest   5–6
self-selection constraint   60–1, 106, 166, 167–8
separating contracts   104–10, 147
  as cost-plus contract   150
  Nash equilibrium   105, 107–8, 119–21, 125

separating equilibrium   98, 110, 121, 124
Shapiro, C.   135
share contract with monitoring   58–9
share price   33, 45
shirking model   130–1, 132–5
shocks   72–3
signalling   32
  compared with screening   117–18
  education for   122–5
social welfare   139–42
Spence, A. M.   126
standard debt contract   8, 32
  and bank competition   61–4
  with monitoring   58, 64
  payoff asymmetry   9–10
state–space representation   85–8
state of the world   5
Stiglitz, J. E.   33, 110–11, 135
subsidy   142
successful outcomes
  probability of   8, 12
  value of   8
Summers, L. H.   135
supply
  and demand interdependency   28
  deposits   13–14, 22, 50
  funds   39
  insurance   82–3
  loans   13–14, 22–4, 50–1, 65

symmetric independent private values (SIPV) model   154, 164, 165

tenders   150
Tirole, J.   151
truth-telling constraints   60–1, 106, 166, 167–8

unemployment   128, 132
unit trusts   18
utility function   143

valuation   154
Vickers, J.   151
Vickrey, W.   155

wages
  efficiency   128–35; optimum   131; reasons for   129–32
  and productivity   115–16
  seniority   131–2
wealth, marginal utility   78–80
Weiss, A.   33
Williamson, S. D.   74
Wilson, C.   111
Wilson equilibrium   109–10, 121
winner's curse   154, 172
Worrall, T.   74

Yarrow, G. K.   151
Yellen, J. L.   135